D0090317

MORE PRAISE FOR *DESIGNING REALITY*

"Bhutan's biggest constraint in promoting Gross National Happiness (GNH), our development philosophy, is its heavy reliance on imports at the end of long supply chains. *Designing Reality* shows that digital fabrication can overcome this constraint by allowing us to fabricate locally while thinking globally and being true to the principles of GNH. We look forward to Bhutan becoming not just a Fab City, but a Fab Country."

　　—Tshering Tobgay, prime minister of Bhutan

"Providing universal access to digital fabrication is one of the most important challenges and opportunities of our time. *Designing Reality* is a manual describing what it is, why it is important, and how to get there."

　　—Congressman Bill Foster, PhD

"In this mind-altering book, the Gershenfelds envision a future of making things that's not dominated by big factories and powerful companies. Instead, it's centered around local innovators using powerful tools to design and build the realities they want. If this sounds good to you, here's the blueprint for making it happen."

　　—Andrew McAfee, scientist, MIT, and coauthor of *The Second Machine Age* and *Machine, Platform, Crowd*

"Ordinary people can now create objects with almost arbitrary levels of complexity, in large part because of the Gershenfelds' insights and leadership. *Designing Reality* is a must-read for anyone who wants to understand this revolution and its implications."

—Erik Brynjolfsson, director of the MIT Initiative on the Digital Economy and coauthor of *The Second Machine Age* and *Machine, Platform, Crowd*

"*Designing Reality* is more than a deep look into the future of making things, it's a sobering (yet entertaining) reflection on how we will need to design society to accommodate the wholesale changes that these technologies are certain to bring. The Gershenfelds have fused their talents to provide a clear picture of how digital materials will come to pass, while addressing the needed transformation in the social sciences if we are to avoid uneven distribution of the benefits. The book offers a highly probable account of a future where error-correcting self-assembly will allow anyone to make (almost) anything."

—James A. Warren, physicist and director of the Materials Genome Program

DESIGNING REALITY

Also by Neil Gershenfeld:

Fab: The Coming Revolution on Your Desktop—from Personal Computers to Personal Fabrication (Basic Books, 2005)

The Physics of Information Technology (Cambridge University Press, 2000)

When Things Start to Think (Henry Holt and Co., 2000)

The Nature of Mathematical Modeling (Cambridge University Press, 1998)

Time Series Preditions: Forcasting the Future and Understanding the Past (with Andeas S. Weingend) (Westview Press, 1993)

Also by Alan Gershenfeld:

Game Plan: The Insider's Guide to Breaking In and Succeeding in the Computer and Video Game Business (with Mark Loparo and Cecilla Barajes) (MacMillan, 2003)

Also by Joel Cutcher-Gershenfeld:

Inside the Ford-UAW Transformation: Pivotal Events in Valuing Work and Delivering Results (with Daniel Brooks and Martin Mulloy) (MIT Press, 2015)

Multinational Human Resource Management and the Law: Common Workplace Problems in Different Legal Environments (with Matt Finkin) (Edward Elgar, 2013)

The Human Side of Enterprise, by Douglas McGregor, annotated edition; updated and annotated by Joel Cutcher-Gershenfeld (McGraw Hill, 2006)

Valuable Disconnects in Organizational Learning Systems: Integrating Bold Visions and Harsh Realities (with Kevin Ford) (Oxford University Press, 2005)

Lean Enterprise Value: Insights from MIT's Lean Aerospace Initiative (with Earll Murman, Tom Allen, Kirkor Bozdogan, Hugh McManus, Debbie Nightingale, Eric Rebentisch, Tom Shields, Fred Stahl, Myles Walton, Joyce Warmkessel, Stanley Weiss, and Sheila Widnall) (Palgrave/Macmillan, 2002)

Knowledge-Driven Work: Unexpected Lessons from Japanese and United States Work Practices (with Michio Nitta, Betty Barrett, Nejib Belhedi, Simon Chow, Takashi Inaba, Iwao Ishino, Wen-Jeng Lin, Michael Moore, William Mothersell, Jennifer Palthe, Shobha Ramanand, Mark Strolle, and Arthur Wheaton) (Oxford University Press, 1998)

Strategic Negotiations: A Theory of Change in Labor-Management Relations (with Richard Walton and Robert McKersie) (Harvard Business School Press, 1994)

ıenfeld
Alan Gershenfeld
Joel Cutcher-Gershenfeld

DESIGNING
REALITY

How to Survive and Thrive in the
Third Digital Revolution

BASIC BOOKS

New York

Basic Books
Hachette Book Group
1290 Avenue of the Americas, New York, NY 10104
www.basicbooks.com

Printed in the United States of America

First Edition: November 2017

Published by Basic Books, an imprint of Perseus Books, LLC, a subsidiary of Hachette Book Group, Inc.

The Hachette Speakers Bureau provides a wide range of authors for speaking events. To find out more, go to www.hachettespeakersbureau.com or call (866) 376–6591.

The publisher is not responsible for websites (or their content) that are not owned by the publisher.

Print book interior design by Amy Quinn

Library of Congress Cataloging-in-Publication Data
Names: Gershenfeld, Neil A., editor. | Gershenfeld, Alan, editor. |
Cutcher-Gershenfeld, Joel, editor.
Title: Designing reality : how to survive and thrive in the third digital
revolution / Neil Gershenfeld, Alan Gershenfeld and Joel
Cutcher-Gershenfeld.
Description: New York : Basic Books, [2017] | Includes bibliographical
references and index.
Identifiers: LCCN 2017023388| ISBN 9780465093472 (hardcover) | ISBN
9780465093489 (e-book)
Subjects: LCSH: Three-dimensional printing. | Three-dimensional
printing—Social aspects. | Manufacturing industries—Forecasting.
Classification: LCC TS171.95 .D47 2017 | DDC 621.9/88—dc23
LC record available at https://lccn.loc.gov/2017023388

LSC-C

10 9 8 7 6 5 4 3 2 1

For our parents, Gladys and Walter,
who designed our reality

N   A   J

CONTENTS

Introduction

Imagine: The year is 1965. Gas is thirty-one cents per gallon. The Beatles have just released the album *Help!* The Watts riots are raging in Los Angeles. *The Sound of Music* is leading at the box office. Digital Equipment Corporation introduces the PDP-8, the first computer to use integrated circuit technology, for eighteen thousand dollars.

You're sitting in a packed coffee shop in San Jose drinking a cup of joe (venti half-sweet no-foam caramel macchiatos have yet to be invented). The only open seats are at your table. A group enters, talking animatedly. They're all holding the latest issue of *Electronics* magazine and are clearly bubbling over with excitement. One of them asks if they can sit at your table. Sure, you say.

You listen in as the group—researchers, it turns out, at a nearby semiconductor company—talk excitedly about an article in the magazine. What they're saying seems completely far-fetched. They're talking about how, one day, computers will be small enough to fit in a pocket or be worn like a watch. How these "personal" computers will be as powerful as a mainframe computer. How, in the near future, all these computers will be connected, enabling anyone, anywhere, to access, manipulate, and share information with anyone, anywhere.

The more they go on, the more their vision of the future sounds fantastical. After all, computers are enormous machines filling entire rooms and run by leading research institutions and big companies. They're expensive and require highly trained operators. The idea that computers could fit in a pocket or be connected seems like something out of Arthur C. Clarke or *Dick Tracy* or *The Jetsons* and not any near-term reality that you need to pay attention to. You can't help but express your skepticism to your tablemates.

The researchers pause and nod as if they've heard this skepticism before. What has made them so excited isn't just that computer technologies are increasing in performance, but that they're doing so at an exponential rate. That kind of change, they explain, can be hard to see in the early phase, before the change becomes so explosive that it is obvious to everyone. And yet, they argue, all the signs of accelerating computing performance are there—if you know where to look. And they insist they know where to look. They're brimming with confidence. You're both skeptical and intrigued. Fine, you say. Where, exactly, should you look?

Encouraged, the researchers begin to explain. First, they say, look at the very nature of digital technology—specifically, what enables digital technologies, unlike most everyday technology, to accelerate at an exponential rate. They compare digital computers to another recent invention, the Xerox machine. Unlike a computer, they explain, a Xerox machine employs an analog process. If, for example, you wanted to make an exponential number of Xerox copies of a document, you'd feed the first copy into the machine and make a copy. You'd then feed the two copies back into the machine to make four copies, then eight, then sixteen, and so on. By the time you get to thousands of copies, let alone a million or a billion of them, most of the copies would be nothing more than a garbled mess because of the accumulated imperfections in the copying process—the information they contained would be lost.

This is not the case with digital technologies, they explain. Digital messages are converted into symbols (ones and zeros). By adding an ongoing process of error correction to the system, a digital machine can double the number of these messages repeatedly without losing any information. This, the researchers explain, is the essence of digital. It enables billions of digital messages to be manipulated and shared with no loss of clarity. Further, the error correction is so cheap that the messages have almost no marginal cost of reproduction. This basic science, they argue, will drive revolutionary changes throughout society.

One of the researchers then opens a copy of *Electronics* magazine and shows you the article they are all excited about. It is written by a man named Gordon Moore, head of research and development at Fairchild Semiconductor, where they work. You're amused at the article's wonderfully direct title: "Cramming More Components onto Integrated Circuits." In the article, Moore makes the case that the number of components on an integrated circuit has been doubling annually and that he expects the trend to continue for at least ten years. The researchers are excited because this trend is essential to computers' ability to become smaller, cheaper, and faster—and it's happening at an accelerating pace. If Moore's prediction is right—if computing performance does continue to double during the next ten years, then by 1975, computers will be a *thousand* times faster. Considering this trend, Moore predicts the development of "such wonders as home computers—or at least terminals connected to a central computer—automatic controls for automobiles, and personal portable communications equipment." He goes on in the article: "The electronic wristwatch needs only a display to be feasible today." You were right to think of Dick Tracy and the Jetsons; that is exactly what Moore is predicting.

The researchers continue to explain. The increasing power of computers makes other digital technologies possible and more powerful as well. For example, one of the researchers is working on a contract from the Department of Defense's Advanced Research Projects Agency to show how computers can communicate over long-distance data networks. Another researcher points out that they are developing computers that will eventually be able to defeat humans at difficult games like chess. These developments, along with many others like them, will intersect and build on each other, laying the foundation for enormous change across every sector of the economy, as well as for culture and society in general.

As your tablemates continue, you become convinced that this is indeed something you need to pay attention to. The exact timing may be unclear, but you realize that you may have just been given a truly unique opportunity to see around the corner into a radically different future. Now you just have to decide what to do with this knowledge.

If you are an entrepreneur, you will probably begin thinking about the potential for new, disruptive business opportunities. If you are the leader of an existing business organization, you are likely to try to use the technologies to gain competitive advantage and to shed or transform parts of your business destined to become obsolete. If you are a leader of a public institution, you might begin to assess how these powerful technologies will reach everyone, not just the fortunate few—how they might be leveraged to enhance the public good and mitigate potential harm. Regardless of who you are, you will consider how these new technologies will have an impact on your personal and professional life. At the very least, it is clear: a digital revolution is coming.

MISSED OPPORTUNITIES

In fact, over the last half century, two digital revolutions have come to pass, more spectacularly than Moore himself predicted. The first digital revolution was in communication, taking us from analog phones to the Internet. The second digital revolution was in computation, bringing us personal computers and smartphones. Together they have fundamentally changed the world.

As early as 1965, the signs of the coming digital revolutions were there for anyone to see. And yet most of the world missed them. As a result, few were prepared for the deep economic, social, and cultural impacts of the first two digital revolutions. Moore's prediction, now known as Moore's Law, didn't just last for the ten years he first projected; it has held for fifty years. And computers are now approximately a *billion* times as powerful as they were in 1965, and they do fit in your pocket and on your wrist.

The technology has continued to advance at an exponential pace, but individuals, organizations, and institutions have largely been playing catch-up, struggling to keep pace. The struggle is at all levels of society—individuals whose lives are increasingly mediated by digital technologies; organizations whose operating models are constantly being disrupted by accelerating technologies; and institutions, such as government, education, and law, that are struggling to maintain stability amid constant change.

The deeper these technologies penetrate society, the greater the struggle for society to keep pace. The best time to shape the destiny of transformative, accelerating technologies is early, before changes have become both widespread and entrenched. This is when the embedded assumptions in the technology and the initial market instantiations are in the early stages of formation and still negotiable. The revolutions in digital communication and computation have enabled unprecedented productivity, generated enormous wealth, and catalyzed remarkable changes in everyday life. But a great many people have also been left behind.

Today, more than a half century after the publication of Gordon Moore's paper, over half the planet still has no Internet access altogether, while billions more have limited or unreliable access. In much of the world, a combination of growing income and wealth inequality, thechnological unemployment, and social polarization, driven by digital echo chambers and "always-on" social media, are ripping at the very fabric of society. Many people feel a deep-seated longing for a simpler, more meaningful and less turbulent future.

The negative aspects of the first two digital revolutions are not simply accidents. Nor were they driven by some unseen hand. Decisions made (and not made) and priorities set (and not set) early on, as the technologies were being developed and introduced to the market, have had lasting effects. We built breakthrough digital communication capabilities, but we failed to build in cultural norms, feedback loops, and algorithms that could have reinforced civil discourse. We created incredibly efficient new models of digital commerce, but have also introduced new threats to privacy and security. We value the advances made possible with digital automation, even as we struggle with the impacts of lost jobs due to technology.

It would have been impossible, of course, to foresee and forestall all the negative consequences of the first two digital revolutions. However, by waiting decades to prioritize helping individuals, organizations, and institutions co-evolve with the technology, we have missed great opportunities to proactively create value and mitigate harm as the technologies were developing. In 2015, President Obama declared that "high-speed Internet is a necessity, not a luxury"—like electricity or water. That was a half century after the publication of Moore's paper. Imagine if developing a culture of digital inclusion, digital and programming literacies, and digital civility was a shared public-private priority starting in the mid-1960s. Imagine if there had been as much social innovation shaping the impact of the technology as there was innovation in the science and technology itself—so that both the social and the technical systems had more effectively co-evolved.

We largely missed this opportunity with the first two digital revolutions. But we now have another chance. It comes with the third digital revolution—in fabrication.

THE THIRD DIGITAL REVOLUTION

The third digital revolution completes the first two revolutions by bringing the programmability of the virtual world of bits into the physical world of atoms. Since that physical world is out here where we live, the implications of the third digital revolution may be even greater than those of its predecessors. This revolution is built on the same fundamental science of digital, only now it enables both bits and atoms to be exponentially manipulated. Just as communications and computation went from analog to digital, resulting in personal computers, mobile phones, and the Internet, the digitization of fabrication offers the promise of personal fabrication, enabling individuals and communities to produce and share products on demand, wherever and whenever something is needed.

The similarities between digital fabrication and digital communication and computation are striking. Much like the early mainframe computers, most serious digital fabrication today is done with enormous machines run by highly trained operators at leading research institutions and big companies. Soon, however, the power that lies in these enormous machines will become accessible to anyone—just as the computer you carry in your pocket has the power of what was once a mainframe computer. We can already see around the corner into a future where anyone can turn data into things and things into data, and can share this information across an Internet of bits *and* atoms.

As with the early stages of the first two digital revolutions, this vision seems, well, fantastical. And yet, not only is this vision theoretically possible, we are already on an exponential path to it becoming a reality. *Fab labs*, community-based labs where individuals can access powerful tools for digital fabrication, have been doubling in number every year and a half, since the first lab was established in 2003. Like the early years of the first two digital revolutions, however, the exponential nature of the third digital revolution is not readily apparent to the casual observer. As inventor and futurist Ray Kurzweil points out in his book *The Singularity Is Near*, "exponential growth is deceptive. It starts out almost imperceptibly and then explodes with unexpected fury—unexpected, that is, if one does not take care to follow its trajectory."

Designing Reality is about taking care to follow the exponential trajectory of digital fabrication. It aims to help you use this knowledge to both prepare for and shape the third digital revolution. In small and possibly big ways, everyone has agency to contribute to this revolution. We do not need to wait a half century for future political, educational, and philanthropic leaders to realize that fab access and literacy is a necessity, not a luxury. We are still in the early stages of the third digital revolution. Research priorities are being formulated, core technologies are developing, and the organizations and institutions essential to universal fab access and literacy are emerging.

Every new technology has inherent attributes that affect the capabilities and behaviors of people using the technology. Digital technologies enable rapid duplication, manipulation, and propagation of content. This property has been central to the transformation of virtually every sector of the economy, how we spend our leisure time, and how we connect with each other. Over the past few decades, we have seen the economic and social impact—both positive and negative—of our ability to duplicate, modify, and share music, videos, blogs, news, email, text messages, and other digital material at virtually no additional marginal cost. This capability is inherent in the very nature of the technology.

Digital fabrication shares some, but not all, of the attributes of digital communication and computation. In the first two digital revolutions, bits changed atoms indirectly (by creating new capabilities and behaviors); in the third digital revolution, the bits will enable people to directly change the atoms. This difference is not yet literally at the atomic level (for most people), but it does mean the ability to use digital design interfaces to modify the physical world. Despite the enormous changes brought on by the first two digital revolutions, much of the physical world around us—roads, houses, appliances, transportation, food—have remained remarkably the same. But in the third digital revolution, the very nature of how the physical world around us is constructed will change. Across the global network of fab labs, we can already see a steady stream of innovations around cost-effective models for individuals and communities to make clothing, furniture, toys, computers, and even houses and cars through designs sourced globally but fabricated locally. These capabilities will continue to improve over time, with exponentially better, faster, and cheaper digital fabrication technologies.

The third digital revolution taps into a deeply held human desire to make things. As Dale Dougherty, founder of *Make* magazine, points out in his book *Free to Make*, "making engages us fully and deeply as human beings, and it satisfies our creative souls," and helps us "see ourselves as confident, capable and creative individuals." Whether it is a machinist at home in a basement workshop, a farmer at a fab lab in rural India, or a twelve-year-old kid at a Maker Faire DIY gathering, making things is deeply satisfying and inspires hobbyists, artists, inventors, engineers, and enthusiasts.

We can see the early signs of the transformative power of digital fabrication in the work of fab pioneers from around the world. Throughout this book, we will meet people like Blair Evans from the Incite Focus fab lab in Detroit. Blair has developed a thirty-acre parcel of land on Detroit's East Side and is boldly exploring new social and economic models where everyone can choose to "work and spend less, create and connect more." Building personal and community self-sufficiency is not a new idea, but Blair and his colleagues are showing how digital fabrication platforms can accelerate self-sufficiency when the platforms are used to collaboratively produce practical goods, from food to furniture.

This increasing ability to make what one consumes at a personal or community level will help address one of the greatest challenges emerging from the first two digital revolutions—the number of jobs being replaced by technology. Some estimates project that as many as half of all jobs could eventually be automated because of advancements in artificial intelligence, robotics, and other rapidly accelerating technologies. These losses are not just blue-collar jobs such as factory work or truck driving, but also white-collar jobs ranging from paralegal work to radiology, and even computer programming. There is debate as to whether new jobs, associated with the new technologies, will make up for lost ones. Even if this emerges (and that is by no means certain), there will be gaps between the loss of old jobs and the emergence of widely accessible new ones. The toxic blend of technologically driven unemployment, income and wealth inequality, and constant change driven by digital technologies, have left many people feeling unmoored and angry, driving nationalist movements throughout the world. These are big challenges, yet the increasing democratization of manufacturing could lead to a more appealing future where personal and community self-sufficiency is combined with global interdependence and knowledge sharing—grounded in capability rather than fear. It could help break down the false dichotomy of globalization and local self-sufficiency, and help transcend political divides.

The foundation for a more sustainable and enriching future is happening not only at a personal and community level, but also on a city and country scale. Another fab pioneer we will meet, Tomas Diez, from the Barcelona fab lab, is leading a global Fab City movement. At the tenth annual global convening of fab labs, the mayor and the chief architect of Barcelona made a bold commitment. They pledged that in forty years, Barcelona would replace its global supply chains with sustainable local production—becoming a city able to make what it needs. This declaration is not a throwback to the craft era, but rather a vision for a postindustrial city where bits that travel globally can manipulate atoms that stay locally to empower cities to become sustainable. Following the Barcelona declaration, several other cities and a few countries, from Santiago to Shenzhen, from Boston to Bhutan, have signed on. Tomas describes the Fab City movement: "We need to reinvent our cities and their relationship to people and nature by re-localizing production, so that cities are generative rather than extractive, restorative rather than destructive, and empowering rather than alienating."

We also introduce fab pioneers working in indigenous communities from Alaska to the Amazon. These innovators are using advanced technologies to keep alive ancient practices for local self-sufficiency and community building. Providing more effective tools to use local materials to collaboratively solve local challenges taps into a deep desire on the part of many people in both traditional and modern societies to be more connected to nature and the physical world around them. Even the most ardent proponents of digital technologies recognize the risk of being sucked into ever-more-enticing social media and immersive virtual worlds. We see anxiety about this risk in the Silicon Valley execs who send their children to "tech-free" schools and in the broader trends such as digital-free holidays and initiatives like the National Day of Unplugging. The third digital revolution can help drive us toward a more healthy balance of time spent in the digital world of bits and the "real" world of atoms.

These aspirational visions will only be urged into existence if we engage and inspire individuals, create the needed organizations, and transform the inherited institutions. The first steps are in recognizing that we are, indeed, on the cusp of a third digital revolution and in understanding its trajectory. The signs are all there, and *Designing Reality* will show you where to look. It will show you how we know the third digital revolution is happening, why it's happening, and, crucially, how you can prepare for and help shape it as it happens.

When exploring exponential technologies, we can very easily fall into either a dystopian vision of the future, where humans have little agency as the technology runs amok and robots steal all our jobs, or be lulled into a techno-utopian vision of the future, where we can just sit back and technology will solve our problems. Both these extremes are continually reinforced in popular media. As we have seen with the first two digital revolutions, the reality is much more textured. The benefits and risks of accelerating technologies are very real, with deep impacts on many lives, but we have the agency, individually and collectively, to shape these impacts now.

We can all help bend the arc of the third digital revolution to create a more self-sufficient, interconnected, and sustainable society. Such a transition will not happen overnight. There will be continued employment for work that cannot be replaced, along with new jobs that leverage these digital fabrication technologies. As individuals and communities increasingly make what they consume, emerging models will challenge our conceptions of work—providing new options for balancing how we will live, learn, work, and play. Making this blend of societal arrangements benefit everyone won't be easy. Yet, if we see a billionfold increase in the capability and reach of digital fabrication technologies over the next few decades, we can realize the vision of working and spending less while creating and connecting more. We could build on the Fab City vision of a society that is "generative rather than extractive, restorative rather than destructive, and empowering rather than alienating."

Designing Reality is written around two broad themes that are together essential for realizing the power and promise of digital fabrication: understanding the technology powering the third digital revolution, and advancing the social systems that must co-evolve with the technology.

TECHNOLOGY

Your guide for understanding the technology is Neil Gershenfeld, director of the Center for Bits and Atoms (CBA) at MIT. Neil has been working at the frontier of digital fabrication for two decades and has a track record of seeing around the corner into the future of digital technologies. His 1999 book, *When Things Start to Think*, anticipated and helped shape what became known as the Internet of Things. His 2005 book *Fab* described the emergence of fab labs and the maker movements, introducing many people to the power and promise of digital fabrication.

Now, in *Designing Reality*, Neil shows how these trends have added up to a third digital revolution that can be seen today after a decade of the spread of the technologies for digital fabrication; how it can be historically understood through the alignment of all three digital revolutions; and how it can be seen in the future in an evolving research roadmap. He shows how the hype surrounding 3D printing is just the tip of a much bigger story, leading up to a *Star Trek*–style replicator that can make (almost) anything, including itself. It will do this by digitizing not just the designs of things but also the construction of the materials that they're made of.

Neil opens in Chapter 1 with the story of the accidental origin of the fab lab movement from his How to Make (almost) Anything class at MIT. He then provides a tour of how community fab labs are being used across the global network, highlighting how today's digital fabrication processes are already empowering people. From the northern tip of Norway to the southern tip of Africa, from rural villages to sprawling cities, community-based fab labs have sparked an outpouring of innovation, providing early indications of the power and potential of digital fabrication.

From the present, Neil then explores in Chapter 3 the history of the core scientific foundations of the third digital revolution. He explains how these go back four billion years to when life evolved the machinery for molecular manufacturing and how the underlying principles of reliability, modularity, locality, and reversibility serve as the core ideas that unify digital communication, computation, and fabrication. He highlights the historical lesson that the implications of these processes could be seen and used long before the technology had reached its final form.

Neil anchors this section with an introduction to Lass' Law, an analogue to Moore's Law for digital fabrication. The law is named for Sherry Lassiter (aka Lass). Along with leading the Fab Foundation (the nonprofit that supports the fab lab network), Lassiter manages outreach for CBA. As the pile of fab lab requests on her desk grew, she first noticed that the number of fab labs was doubling roughly every year and a half.

When Neil wrote *Fab*, he neither planned for nor envisioned this exponential growth. At the time, CBA had only recently launched the first few fab labs. For Neil, the year 2003 was the equivalent of 1959, the start of the data points that Gordon Moore had plotted. Now, after more than a decade of the number of fab labs doubling—along with continued advancements in the research roadmap and a growing fab ecosystem—we can predict a likely exponential trajectory for digital fabrication performance and reach.

As with Moore's Law after its first decade, we can extrapolate this exponential trend. This means the equivalent capability of a million fab labs over the next ten years or so, and a billion over the following ten years. This doesn't mean a billion room-filling facilities. Rather, each of these decades marks a technical stage in the integration, accessibility and reach of the technology. As the scaling progresses, what is counted is no longer fab labs as they exist today but rather the equivalent capability to fabricate physical forms and program their functions. As Moore's Law reaches its physical and economic limits, Lass' Law continues an exponential growth trend.

This brings us to Neil's Chapter 5, a roadmap for the future of digital fabrication. He opens the chapter with an inventory and a description of the current tools in a fab lab. He then outlines four distinct stages: community fabrication (powered by computers controlling machines), personal fabrication (based on machines that can make machines), universal fabrication (marking a transition to digital materials), and ubiquitous fabrication (with programmable materials). Each of these stages represents an exponential improvement in digital fabrication performance.

Much as the core elements of the first two digital revolutions were visible in the labs of the mid-1960s, when Gordon Moore wrote his article, all the core elements of the third digital revolution are visible in research labs today. The question is, how long it will take for them to emerge from the lab and impact society? And will we be ready?

SOCIETY

Your guides for exploring how social systems can effectively co-evolve with the accelerating technologies of the third digital revolution are Joel Cutcher-Gershenfeld and Alan Gershenfeld. Joel is a professor in the Heller School for Social Policy and Management at Brandeis University and past president of the Labor and Employment Relations Association. He has researched and facilitated large-scale systems change in the auto industry, the aerospace industry, health care, biomedicine, and the nonprofit sector. Joel is a macro social scientist with a track record for advancing theory and practice in negotiations and high-performance work systems. He is now pioneering new models for multi-stakeholder alignment within and across levels: workplace, enterprise, community, industry, national, and international levels.

Alan is president and co-founder of E-Line Media, a "double-bottom-line" company (committed to both positive financial returns and meaningful social impact), harnessing the power of digital media and games to help people understand and shape their world. In his work at E-Line and in his former roles as a studio head at Activision and board chair for Games for Change, he has worked on social-impact media projects with NSF, DARPA, USAID, the White House Office of Science and Technology Policy, the Smithsonian, PBS, the Gates Foundation, the MacArthur Foundation, and others that have collectively engaged and empowered millions of people all over the globe.

Both Joel and Alan have been collaborating with Neil in the fab lab movement since the launch of the first lab, where Joel and his oldest son volunteered for many years. Joel also helped launch a fab lab in Champaign-Urbana, Illinois, and has led the application of new stakeholder alignment methods across the fab network. Alan has researched sustainable business models for fab labs, and E-Line has worked on a DARPA-funded video game exploring the future of digital fabrication, in collaboration with the Fab Foundation and CBA. In conducting research for this book, Joel and Alan have visited fab labs throughout the world, interviewed dozens of fab pioneers, and surveyed hundreds of stakeholders.

Neil explores the technology roadmap and offers tools and techniques for leveraging the technology; Joel and Alan explore the social roadmap and offer methods and mind-sets so the social systems can co-evolve with the technical systems. In Chapter 2 Joel and Alan also open by observing the current global network of fab labs. But unlike Neil, they don't only present exhilaration and empowerment; they also highlight the tensions and challenges that permeate the fab network. Despite the promise of personal fabrication and individuals' ability to make what they consume, digital fabrication is still a long way from being a reality for most people. There are significant challenges around fab access, literacy, and the cultivation of an ecosystem that ensures truly democratized technologies. A close look at these challenges is essential if we are to address them. Observing disconnects throughout the fab ecosystem provides a window into the embedded underlying assumptions and conflicting values. Today, the benefits of fab labs come more from the process of making than they come from the result, because of the difficulty of mastering current workflows. Over time, this balance will need to change if the impact of digital fabrication is to move beyond just the cultivation of new skills and dispositions to increasingly make what we consume.

Like Neil, Alan and Joel then turn to the past. In Chapter 4, they highlight how Moore's Law is as much a social construct as a technical one. Moore's Law was never a law of physics, but rather an observation that became a core business strategy for a company, an industry benchmark, and finally a galvanizing framework for better, faster, cheaper digital technologies. The chapter illustrates how the interweaving of social and technical change is not new—how the modern social sciences, that is, the study of human societies and relationships, began in reaction to the rise of the industrial revolution. And yet, because the social sciences began by reacting to technology, the dominant practice in this field is the observation of technology rather than its co-creation. Taking a more proactive stance begins with understanding rates of change for individuals, organizations, and institutions, which takes on new meaning in a world of accelerating technologies. This points to the key roles of rate limiters and rate accelerators to positive social change, as well as lessons learned from the first two digital revolutions in harnessing the power of digital platforms and emergent ecosystems to effectively co-evolve technical and social systems.

Joel and Alan conclude with specific guidance for shaping the future of digital fabrication (Chapter 6). They describe eight aspirational scenarios, jointly created with fab pioneers across five continents. In these scenarios, social systems and technical systems co-evolve in transformational ways. To shape the future, we need mental maps for not just possible futures, but for preferable ones. To help transform these preferable futures into reality, Joel and Alan offer a framework for fostering new mind-sets and applying effective methods so that everyone can find meaning, purpose, and dignity in the third digital revolution.

Designing Reality brings the perspectives of science, technology, social science, and humanities to the third digital revolution—through three brothers who are not only observers of the revolution but also active participants in helping guide it. Each brother brings a different lens to the book. Like all lenses, each brother's clarifies some things and makes other things less clear. Their disagreements have been even more important than their agreements as they came together to write this book. The same is true any time very different sectors or domains need to collaborate around complex, rapidly changing technology to accomplish collective goals—collaboration that the third digital revolution demands.

This is a unique historical moment: we can foresee the likely trajectories, and it is still early enough that we can shape the technology before it shapes us in ways we will regret. The stakes are high. If our projection is correct, the third digital revolution will have as much impact as, if not more impact than, the first two digital revolutions. We will soon be facing a torrent of new opportunities and challenges that go to the very heart of how we exist. Literacy scholar James Paul Gee from Arizona State University summarizes the opportunity and challenge in his essay "Literacy: From Writing to Fabbing":

> Fab could create a world with yet deeper inequalities than we currently have, a world where only a few engage in the alchemy of turning ideas into bits into atoms and back again. The rest will live in a world where the stuff of life and the world—objects, cells, materials—are owned and operated by only a few. Fab is a new literacy and we have as yet no real idea how it will work out. But it is a special and, in some sense, final one. . . .
>
> How many of us will get to be homo fabber? Humans have always been the ultimate tool makers. Soon the tools for world making will be cheap enough to be in the hands of everyone, should we want to make that happen. Will we, as a species, make a better world or a worse one when some or many or all of us become god-like creators, calling worlds into being? Fab is to literacy what fire was to human development: a tool that can light the way or burn it down.

Will we light the way or burn it down? We have the agency, individually and collectively, to tip this balance. As digital fabrication becomes increasingly democratized, we'll have the ability to leverage bits to manipulate atoms to improve lives. We will be able to design reality, both metaphorically and literally.

CHAPTER 1

How to Make (almost) Anything

The first digital revolution was in communication. Before that, analog telephone calls degraded with distance. We now have a globe-spanning Internet that makes it as easy to talk to someone around the world as it is to chat with someone around the corner.

The second digital revolution was in computation. Analog computers used to fill rooms with gears and pulleys or vacuum tubes and produced answers that accumulated errors the longer that they ran. Today, you can carry in your pocket a computer with the power of what was once a national lab's supercomputer.

We are now living through the third digital revolution, in fabrication. The first two revolutions rapidly expanded access to communication and computation; this one will allow anyone to make (almost) anything. This time around, it's likely to be even more significant than the first two, because it's bringing the programmability of the world of bits out into the world of atoms.

The defining application for digital computing was personal computing, which upended the existing computing industry that initially ignored it. Likewise, the defining application emerging for digital fabrication is personal fabrication, which allows consumers to become creators, locally producing rather than purchasing mass-manufactured products.

Digital fabrication has a decades-old meaning, referring to computers controlling machines that make things. And it has a much deeper meaning that, as we'll see in the coming chapters, is both much newer and much older: the digitization of not just the description but also the actual construction of an object. As was the case with the earlier digital revolutions, we don't need to wait for the technology to reach its final form to recognize or use it. The third digital revolution can be seen today in the spread of technology for digital fabrication and the impact that it is already having (the subject of this chapter). It can be seen in the historical alignment of all three digital revolutions (Chapter 3). And it can be seen in the coming research roadmap (Chapter 5). Together, these chapters survey the science and technology required to understand the third digital revolution, providing the background needed to be able to shape it.

The exponential change in all three digital revolutions began with weak signals in the first few doublings; today, the signals for the third digital revolution are more like honking horns (if you're paying attention to them). This chapter examines what is already happening today, introducing Sherry Lassiter's original observation (which we're calling Lass' Law) that the number of fab labs has been doubling every year and a half. It explains what a fab lab is, how to use one, what the applications and implications of these labs are, and how they are organized.

This tour through the present is important for seeing the future because the most significant implication of projecting the continuation of Lass' Law, like Moore's Law, is to change not just what the technology can do, but who can do it. On this tour, we'll meet pioneers who are already using fab labs to produce a range of remarkable things, fabricating the physical forms of objects and programming the functions that the objects can perform. Today, these tasks require access to the tools in a fab lab along with the supply chain that supports one. Over time, the equivalent capabilities will become available to many more people as the progression of performance improvements we'll see in the subsequent chapters decreases the cost and increases the capability of digital fabrication.

FABRICATION

Fab labs are laboratories for fabrication (which we also think are fabulous labs). They began as an outreach project from MIT's Center for Bits and Atoms (CBA), which I direct. CBA was founded to study the boundary between computer science and physical science, a distinction that I never understood. Computation both requires and is used to represent physics; CBA researchers have participated in projects like creating among the first computers to use the strange behavior of microscopic quantum systems to solve important problems faster than a conventional classical computer can. Another CBA collaboration contributed to creating some of the first

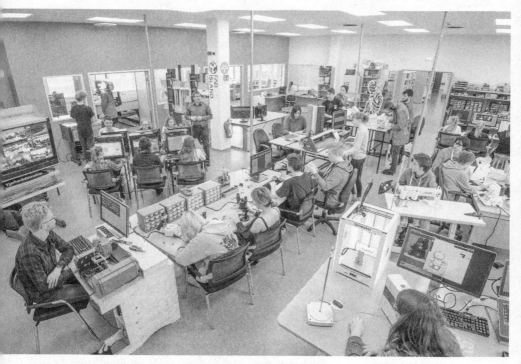

A fab lab in Vestmannaeyjar, Iceland. *Frosti Gíslason/Saethor Vido*

living organisms designed in a computer. These activities can't be neatly separated into hardware and software; they are intimately integrated. The most important conclusion from this research is the recognition of the fundamental convergence of digital communication and computation with fabrication, a concept that will be unpacked in the coming chapters.

The initial funding to create CBA came from the National Science Foundation (NSF), which supported our ambitious proposal to put together a facility that could make and measure things with sizes ranging from molecules to buildings. CBA includes million-dollar research instruments like electron microscopes and X-ray scanners. Within that facility is a workshop containing hundred-thousand-dollar manufacturing tools, like machining centers, that we use to develop the experimental apparatus. Within the workshop is a collection of ten-thousand-dollar rapid-prototyping tools, like laser cutters, that make parts of projects. And those in turn get used with thousand-dollar tools to perform processes, like molding components. The nesting of these scales is key to understanding fab labs, which fall in the middle of this hierarchy; they represent the core capabilities needed to make not just something but (almost) anything.

Fab labs began as an experiment to see what would happen if the most popular of CBA's tools internally become widely available externally. They arose not from a transformative vision but much more modestly to meet a bureaucratic requirement. When CBA started in 2001, the government had begun enforcing legislation called the Government Performance and Results Act, which required agencies to measure their progress against performance goals and to document their impact in a broader context. NSF responded in turn by asking grantees like CBA to show the broader impacts of their work. My colleagues and I, who had no idea how to do that, had encountered an unexpected impact in rural India.

On a trip there in 2002, I met S. S. Kalbag, who had run research for Hindustan Lever. When he reached the Hindu life stage when it's traditional to renounce worldly attachments, he approached that as a scientist by setting up a school (Vigyan Ashram) to teach technical skills to dropouts. He intentionally located this in one of the poorest, driest parts of India: Pabal, in western Maharashtra. When I visited him there, he had a long list of local needs that he could have met with the tools in his former lab but not with the resources available in the small village of Pabal. Rather than invest in expensive special-purpose lab equipment, we began a collaboration to equip their lab to make lab instruments for purposes like agricultural testing. That could be called fab lab number zero, the chicken before the egg (or is it the egg before the chicken?). It inspired the first full community fab lab that was later opened in Boston, and itself grew into a full fab lab in 2005. Although Kalbag passed away in 2003, that fab lab has flourished under his successor, Yogesh Kulkarni, teaching classes, incubating businesses, and supporting the community.

To respond to NSF's requirement to show broader impacts, we proposed to base an outreach program on the experiment in Pabal. Our adventurous program managers agreed, and the concept of a fab lab was born. The contents of a fab lab were based on a kind of market research done by running CBA's facility at MIT, incorporating the most useful core set of tools. Today, this suite adds up to about a hundred thousand dollars, weighs around two tons, and fills a room. A fab lab naturally includes a 3D printer, which has been the subject of a great deal of coverage in the popular press, but it is just one of the computer-controlled machines. We'll meet the rest in Chapter 5, including a laser cutter that's much faster than the 3D printer, a large milling machine that can make things like furniture, a small precision milling machine that can make electronic circuit boards and molds for casting parts, tools to assemble and program electronics, a scanner to digitize objects, and computers for design and modeling.

When microwave ovens were introduced, they were the basis for the 1950s version of the push-button kitchen of the future. You, of course, still use a stove, an oven, and maybe a toaster along with a microwave. All these tools just heat food, but each is needed to make a range of recipes. Working in a fab lab today is like cooking in a kitchen. Think of the 3D printer as the microwave oven of the fab lab. You could use only the microwave, but you would be missing the capabilities of the other appliances. Just as a basic set of processes is assumed in cooking, a basic set is assumed in digital fabrication.

You can effectively consider the whole fab lab to be a machine: data goes in and things go out, things go in and data goes out. What will change over time is not what can be made but rather what's required to make something. A fab lab today uses bulk materials that can be locally sourced, like wood or cardboard, along with a small set of globally sourced high-tech consumables, like precision bearings and computer chips. The latter can't yet be made in the lab, but that will become possible in the coming years. The transition will be continuous, as more and more of the supply chain to support a fab lab gets replaced with fewer and fewer inputs.

When we were planning to set up the first of these fab labs, Bill Mitchell, then MIT's thoughtful dean of architecture, suggested that I talk to Mel King in Boston. Mel is a community activist who literally helped invent mixed-use urban development by the seat of his pants. He led what was called the Tent City encampment, which forced a developer to include affordable housing and community space in a planned parking garage. Mel's South End Technology Center (SETC) in the resulting complex went on to become a pioneer in electronic media at a time when mass media wasn't telling stories of the inner city. Then SETC innovated in computing access when the Internet threatened to bypass the community. So it was natural to progress from digital communications and computing to fabrication, opening a fab lab there in 2003. As Mel says, "the rear wheels of the train don't catch up to the front wheels of the train unless something dramatic happens to the train," meaning that each of these interventions was a disruptive event that challenged the role of technology in society.

Putting a fab lab at SETC was the extent of our vision. After it opened, we expected to just return to research. But a strong Ghanaian community in Boston, after seeing Mel's lab, collaborated to bring a fab lab to Sekondi-Takoradi on Ghana's coast in 2004. From one to two, then four, the number of fab labs has continued to double every year and a half for a decade: every time we opened a fab lab, someone else wanted one. Sherry Lassiter was the first to notice this exponential trend; she's managed the fab lab program for CBA, and leads the Fab Foundation that was spun off to support its growth. Lass came to me with a background in producing science programs on television. She was interested in producing the science itself, which is what she has done ever since with the fab lab network.

In 2005, I wrote the book *Fab* after the first few doublings of fab labs. That year, I was approached to be the first interview in a new magazine called *Make*, which was founded by Dale Dougherty. He coined the term *maker* to describe the emerging community of hobbyists connecting computation with fabrication using the kinds of tools found in a fab lab. He started hosting gatherings called Maker Faires in 2006; the biggest of these, in San Mateo, California, grew from a modest beginning to attract 145,000 visitors in 2015.

Also in 2006, Jim Newton founded TechShop to provide shared access to digital fabrication tools beyond the reach of most individuals. These shops are run with a paid membership model; as of 2016, there are ten of these. More informally, maker spaces and hacker spaces started to spread to provide a place for like-minded individuals to gather. The spaces vary widely in what they offer, but they now number in the thousands.

The next year, in 2007, CBA launched a mobile fab lab to bring tools to people rather than vice versa. Tomas Diez, Amy Sun, and Kenny Cheung (whom we'll meet later) memorably commissioned the lab by driving it to the Burning Man gathering in Nevada's Black Rock Desert, for rapid prototyping on the playa. The mobile lab subsequently seeded a network within the network: roving fab labs touring regions of the country.

All these fab labs were early manifestations of the third digital revolution. Two things distinguish fab labs within this growing ecosystem. First, rather than each one being different, fab labs all share the same evolving set of core capabilities, allowing people and projects to be shared among them. The computer networking pioneer Bob Metcalfe observed what is now known as Metcalfe's Law: the value of a computer connected to the Internet is proportional to the square of the number of computers in the network. He proposed that it's the square, rather than merely the number of computers, because that's how many pairs can talk to each other. Some kinds of maker spaces are based on a membership model that's like joining a gym. Gyms provide individual access to exercise equipment; there's no direct benefit if someone is exercising elsewhere. But the value of being connected to the Internet, or working in a fab lab, increases when other computers are connected to the Internet, or when other people are working in fab labs. Both people and projects are mobile in the fab lab network, sharing content that allows them to accomplish collectively what they couldn't do individually.

The other distinguishing feature of fab labs is the coordinated evolution of their contents according to the digital fabrication research roadmap presented in Chapter 5. They began with a carefully curated inventory of common machines, materials, components, and programs and are now migrating to open designs for hardware and software developed by and for fab labs toward the goal of a fab lab's being able to make another fab lab. Although the cost of each type of machine in a fab lab has come down over time, the ambition of what can be made in a fab lab has gone up at the same rate, so the overall cost has stayed roughly constant, on the scale of a community resource.

Together, these attributes allow the collection of fab labs to function as a network. Individually, each site isn't a critical mass, but the collection of them is. No one is pushing anyone to start a fab lab, but sites continue to join the network for the benefits they get from being part of something larger.

The biggest surprise for me has been how similar rather than how different the uses of fab labs are around the world. Mel King captured this when we took him, a community activist from Boston, to the far north of Norway to meet the Sami-descended herder Haakon Karlsen. After spending a few days in Haakon's fab lab, Mel commented that it was "just around the corner." He might have been a few hours above the Arctic Circle, but he recognized the same hopes and fears as those he found in his own urban lab.

Common to fab labs is how they mix ages, from very young to very old, and applications, spanning education, entertainment, and business. In this diversity, they're serving a role analogous to libraries. Andrew Carnegie invested in setting up town libraries around the turn of the last century (1900); by the time he was done, there were about twenty-five hundred such libraries dotting the nation. The overall mission of a library is literacy, expanding access to knowledge. But within that mission, they're used for purposes ranging from playgroups to classes to research to civics. From what was initially a novelty, libraries are now an expected component of a civilized community. You can think of fab labs as doing the same, but they now aim at a new form of literacy: going from bits to atoms.

EDUCATION

Once CBA had set up its digital fabrication research facility, we had a problem: because these tools are conventionally segregated by both discipline and the scale at which they operate, it would have taken a lifetime of MIT classes to learn how to use them all. As a shortcut, in 2001, I started a new class: How to Make (almost) Anything. The class was aimed at a small group of students doing digital fabrication research, but every year since then, hundreds of students have shown up for the class, just wanting to learn how to make things.

Along with mastering individual skills, they did projects to integrate these skills. One of the stars the first year was Kelly Dobson, who went on to become the head of the Digital + Media department at the Rhode Island School of Design. She made a wearable device that could save up screams and play them back later when it was convenient to let them out. A few years later, Meejin Yoon, who later became the head of the department of architecture at MIT, made a dress, replete with sensors and spines, that could defend a wearer's personal space. These kinds of inventive projects happened so consistently year after year, I realized that the students in the class were answering a question that I hadn't asked: What is digital fabrication good for? While I was asking how to do it, and not why, they

Hans-Kristian Bruvold (*left*) and Tshepiso Monaheng working in their fab labs. *Neil Gershenfeld*

were showing that just as the killer app for digital computing was personal computing, the killer app for digital fabrication is personal fabrication. The point was not to make what you could buy in stores; it was to make what you couldn't—products for a market as small as one person.

In the same way that the arrival of CBA's research tools presented a training problem that was solved by the How to Make (almost) Anything class, the spread of fab labs made training a problem on a global scale. Bright kids would learn skills in fab labs that were far ahead of local educational opportunity, and then they'd fall off a cliff. Hans-Kristian Bruvold was considered something of a problem in the local school system of Lyngseidet in the far north of Norway. Because he had already mastered everything the teachers were teaching, he wasn't an attentive student. He started going to Haakon's fab lab instead, which is where I met him and showed him a few demonstration projects from my How to Make (almost) Anything class. When I next returned, I was astounded to see that he had integrated the techniques into a toy robotic truck, including the design of the body, incorporating the motors and their controllers, and adding a windshield display. In South Africa, something similar happened when we opened a fab lab in what had been an apartheid-era township, Soshanguve. There, I was startled to later find that a local girl, Tshepiso Monaheng, had been using the lab to remotely follow along with the work of my classes at MIT.

The usual message for someone like Hans-Kristian or Tshepiso is, "You're smart, so you have to leave now." Bright students like them have to go far away to study somewhere more advanced. But this migration takes the most valuable people away from where they're most needed. We initially tried to pair with local schools around the world to fill this void but consistently found that an even greater limitation than a lack of technical skills was how a school's regimented approach to education can stifle creativity. For these reasons, we started what's now called the Fab Academy.

It grew out of a video link that I initially had set up so that fab labs could remotely sit in on the How to Make (almost) Anything classes at MIT. When there were more fab labs attending than students in person, we spun off the remote sessions as a separate program. The local mentors, who had initially been the remote students, proved to be essential. New students joined workgroups in their local fab labs, where they worked with these mentors, their own peers, and the machines. We then connected everyone globally by video for interactive lectures and collaborative content sharing.

In computing terms, you can think of MIT as a mainframe where you go for processing. It works well but for a very limited population. You can think of massive open online classes (MOOCs) as corresponding to the time-sharing era in computing, when users sat at isolated terminals connected to central mainframes. And the Fab Academy model that we stumbled on is more like the Internet, linking nodes in a learning network that grows at its edges rather than at its center.

Initially I directly supervised all the students. Then as the model grew, I supervised the mentors who supervised the students. As it grew further still, I supervised supernodes that emerged to supervise regional labs, which in turn supervise the students. In this, the model is again like the Internet, which handles the routing of information in a tree with trunks, branches, and leaves. And like the Internet, any node can talk to any other; the heart of the weekly Fab Academy cycle is a giant video conference where everyone can see and hear everyone else. The conference includes a lively discussion of successes and failures in the preceding week and an interactive introduction to new material for the next week. All this collaboration is supported by a distributed workgroup led by Luciana Asinari rather than a central office.

This structure maintains a direct traceability in the web of relationships to maintain quality control. But we needed a way to document that, which led me to approach EDUCAUSE, the group of IT professionals in higher education that runs the .edu domain. They require institutions wanting an .edu domain for their websites to be accredited. The accreditors that I spoke with appreciated what we were doing but explained that giving the Fab Academy an .edu domain name would violate their rules. Because they accredit organizations that have a physical place, the accreditors have no way to recognize a network. But the accreditors then said something helpful: "Pretend." By that, they meant we should have students build portfolios documenting the skills they're learning, and the group would eventually catch up to us to recognize the students' work and our evaluation. Although the Fab Academy has no global accreditation, regional accreditations are now beginning to be overlaid on the diploma that the Fab Academy awards. And we've found that for future admission, employment, or investment, the portfolios can matter more than a credential from an unknown body. The cycle of content and evaluation takes about eight months to cover, but a student's progress is determined by the mastery of the skills rather than time in class—some students have taken a few years to finish everything. The level of the students has ranged from home-schooled prodigies to college students, to people doing this instead of college, to midcareer professionals, to late-career retraining, to retirement avocations. This global linking of local learning workgroups balances the distributed nature of fab labs with the need for mentoring.

Absent good mentoring, bad ideas can propagate. The term *maker* has come to have both a negative and a positive connotation, along the lines of *enthusiastic but not well informed*. A staple of the maker movement is the Arduino, a twenty-dollar small computer board used to build intelligence into projects, including reading sensors, controlling output devices, and communicating with networks. The Arduino originally in turn was based on a computer chip family called AVRs, which were designed by two Norwegian students. After introducing the Arduino, the Fab Academy shows how to make such a board in a fab lab for a few dollars in parts. Students then learn how to use other computer chips, from the size of a rice grain up to something that can run a desktop operating system. Another staple of the maker movement is 3D printers. After showing students how to use one, the Fab Academy shows them how to use all the other digital fabrication tools that can operate more quickly or make larger things, stronger things, or things with finer features. Then students learn how to make a 3D printer. These examples each provide a path from introducing easy skills to mastering hard ones.

Do-it-yourself is a recipe for standing on the toes rather than the shoulders of your predecessors; do-it-together or do-it-with-others builds on their accumulated knowledge. We unexpectedly found the Fab Academy to be at the heart of a virtuous circle. Each cycle would propagate best practices throughout the fab lab network, building a core collaborating community of local mentors and providing a cohort of trained students that then became available to help with new labs and programs.

The Fab Academy was developed to teach digital fabrication. But much of what we had assembled wasn't specific to that content; the infrastructure could be used for any kind of distributed rather than distance learning. I had initially missed the deep connection between communication, computation, fabrication, and learning. Whereas digital communication lets us interact globally and digital computation lets us share knowledge, the addition of digital fabrication lets us exchange things as well as ideas. With the core set of tools in a fab lab, it's then possible to make whatever else is needed, effectively bringing the campus to the student.

George Church, one of the world's leading geneticists, was interested in reaching students beyond those who could fit into his classes at

Harvard. This thought led George to add a second distributed class in the fab lab network: How to Grow (almost) Anything. Digital fabrication and biological fabrication connect at two levels. Biologists can use a fab lab to make the tools needed in a bio lab; biological equipment is often both overpriced and cumbersome. The same techniques used to make machines in fab labs have been used to make things like thermal cyclers for DNA amplification and liquid-handling robots to program reactions. At a deeper level, biology itself can be used for fabrication. As we'll see in Chapter 3, biological processes are fundamentally digital, and we're increasingly learning how to program these processes, as fab labs and bio labs converge.

Olafur Eliasson is one of the world's foremost artists. Like George, he wanted to extend his influence beyond the students he could directly teach in his studio. But his interest was not in *how* to make things, but *why*, leading him to begin developing another distributed class, Why Make (almost) Anything, to explore the influences on, and impacts of, the making process.

My student Nadya Peek jokingly called this growing collection of programs the Academy of (almost) Anything. The name stuck, or "Academany" for short, and it is now managed by Jean-Michel Molenaar (who started the Grenoble fab lab). Each of its offerings follows the same model of local workgroups, with mentors connected globally for interactive lectures from world leaders with collaborative content sharing.

While fab labs were spreading around the world, I helped plan a new building at MIT. The task took ten years from start to finish, cost a hundred million dollars, and fits a few hundred people. Each of the thousand fab labs that emerged over those ten years has a community of a hundred or so users. These numbers pose an obvious question: Which activities justify the hundred-million-dollar versus the hundred-thousand-dollar investment?

The existing organization of MIT is based on an assumption of scarcity. To manage access to our tools in labs, books in libraries, and faculty time, we reject most applicants and crowd into a corner of Cambridge, where there's a battle over every square foot of space. It's a false dichotomy to consider the alternative an isolated student sitting in front of a computer connected to an online learning platform; we've consistently found in the Fab Academy that for students to succeed, they need to be in learning communities. The real alternative is distributed rather than distance education, as the Fab Academy backed into doing. The follow-up question is then, how much of what is done at a place like MIT can be distributed this way, and how much needs to be centralized?

I'd say about half. Whenever we open a fab lab, we find the same kind of remarkable, inventive people who I work with at MIT. They're everywhere, appearing so consistently in fab labs because they're unable to find peers, mentors, and tools. About half the activities on MIT's campus could be done in a fab lab setting. The other half differ in that they require much more expensive tools, like the nanoscience instruments we're using to develop molecular-scale assemblers. The skills and knowledge to use these expensive tools are so specialized that it makes sense to do these activities all in one place. These two types of spaces aren't in opposition—you can view all this as a tree, with ten-thousand-dollar maker spaces, hundred-thousand-dollar fab labs, million-dollar super fab labs, and ten-million-dollar research labs. But it's by growing the tree out rather than up that we scale to tap the brainpower of the planet.

Seymour Papert is the father of computers and education. He studied in Switzerland with the pioneering child psychologist Jean Piaget, who argued that children learn like scientists, by doing experiments and testing theories. Seymour then came to MIT to get access to early real-time digital computers, wanting to expand the scope of experimentation available to a child. This was an improbable thought at the time—these computers were expensive, room-filling beasts that were difficult to use. To provide a friendlier interface, Seymour developed robotic "turtles" that he connected to the computer, and a language (Logo) that let children tell the turtles what to do.

One of the people who came to work with Seymour is Alan Kay, who went on to develop the modern computing paradigms of graphical user interfaces and laptops. These design principles weren't originally intended for business executives to balance spreadsheets; they were for children to discover. Another person who studied with Seymour is Mitch Resnick, who developed Lego's Mindstorms kits (named after a book that Seymour wrote), which moved the computer into a programmable Lego brick. Mitch also led the creation of the popular Scratch software for kids to program.

As fab labs started doubling and the Fab Academy began to grow, Seymour came by to see me to talk about them. I had considered the whole fab-lab thing to be a historical accident, but he made a gesture of poking his side. He said that it had been a thorn in his side that kids could program the motion of the turtle but could not make the turtle itself. This had been his goal all along. Viewed that way, learning in fab labs follows directly from the work he started decades ago. It's not an accident; there's a natural progression from going to MIT to play with a central computer, to going to a store to purchase and play with a toy containing a computer, to going to a fab lab to play with creating a computer.

APPLICATION

Once you're equipped with access to both the tools in a fab lab and the ability to use them, it's possible to locally produce the kinds of products that are today purchased at the end of long supply chains. Along with the benefits of using local skills and creating local jobs, a fab lab allows on-demand production and customization to meet local needs. Here are examples of how fab labs are being used, a survey that's intended to be illustrative but not exhaustive (or exhausting).

Craft

The Cook Inlet Tribal Council (CITC) is a tribal nonprofit serving Alaska's Cook Inlet region. The Alaska Native communities that it serves have profound cultural traditions but also serious issues with unemployment, alcoholism, and suicide rates. CITC hosts a fab lab that opened in 2013 and is focused on merging culture and technology to serve a new generation that is growing up surrounded by digital devices that frequently have little local context. The White House's 2014 Native Youth Report found that the high school graduation rate among Native high school students is the lowest of any demographic group across all schools.

Benjamin Hunter-Francis II was sixteen and at risk of becoming one of those statistics when he moved to Anchorage from the remote village of Marshall, Alaska, population 349. Far behind in school, he had a different sense of culture from kids born and raised in the city. He became a fixture in the fab lab, using it both for his classes and for personal projects. One project was a wooden sled based on traditional designs but made with computerized tools and engraved with images from his community. For another project, he used the laser cutter to do marquetry that's traditionally done laboriously by hand carving. Along with catching up and now enjoying school, he feels that the fab lab is helping him keep in touch with his ancestors, traditions that he wants to help keep alive.

Haystack Mountain School of Crafts is one of the premier artists' colonies in the United States. Here, renowned glassblowers, blacksmiths,

Fab lab crafts from (*clockwise in pairs from the upper left*) Alaska, Mexico, Maine, and Japan. Benjamin Hunter-Francis II, Gonzalo Pérez Ramírez, Andrea Dezsö, Hiroya Tanaka

potters, printmakers, and other artists retreat to this collection of studios on the coast of Maine to practice and teach their crafts. In 2009, after the then director Stu Kestenbaum proposed as an experiment to introduce digital fabrication tools, we set up a temporary fab lab there. The response was a bit like when Bob Dylan showed up at the Newport Folk Festival in 1965 with an electric guitar, an event that one observer said "electrified one half of the audience, and electrocuted the other." Half of the artists were horrified by the intrusion of technology into a temple of traditional crafts; the other half were horrified by the first half for not recognizing that all their practices rested on technologies that were once new and that this was just another one.

So many artists had life-changing experiences in the fab lab that in 2011 it became a permanent addition and the only controversy was contention for time on the laser cutter. But rather than being viewed as one of the crafts, the fab lab tools were used across all of them. And rather than replacing an artist's skills, the lab was used to amplify them. Artists would typically design in traditional media, then use the fab lab to embody their work in ways that would be difficult to do by hand.

Andrea Dezsö was a visiting visual artist who makes gorgeous, intricate paper cutouts and prints. She starts with sketches of these, and then a laborious process of cutting follows. In the fab lab, she could scan her sketches and rapidly turn them into objects, with features that were both larger and finer than she could have done by hand. Most interesting for her was the collaboration with the machines. They didn't always do what she expected; sometimes the results were better than what she had intended. For example, while cutting out a design with a tool path that unintentionally moved in steps larger than the tool, she found that the material left behind created what looked like an evocative energy field around the figures.

Universidad Anáhuac México Norte opened the first fab lab in Mexico, in 2012. It has a strong focus on social impact, empowering marginalized women who make Mexican crafts. One of its first projects was with a single mother, her six daughters, and several of their cousins—the Arreola family. Although they were underemployed, the Arreolas knew how to make handmade chocolate. But the presentation was crude, limiting their ability to sell it. They were able to use the fab lab to make molds to form the chocolate and to produce custom packaging to present it. The result was a much higher-quality product that considerably increased their sales while also saving time and effort.

Fab Lab Kamakura, founded by Hiroya Tanaka and Youka Watanabe in 2011, occupies a 150-year-old Japanese wooden *kura* building that had been used as a sake brewery. This lab focuses on connecting traditional Japanese crafts with modern tools. One project developed the use of Japanese urushi lacquer to coat the output from 3D printers for a beautiful, tough, glossy finish. Another studied the construction of the building itself. One of the students working in the lab, Kenji Kanasaki, identified fifty-four types of joints in traditional Japanese woodworking, far beyond the simple tabbed joints commonly used (and overused) in fab labs. He turned these into a set of reference design files that could be made with the tools in a fab lab, creating a dialogue between the tradition of Japanese joinery and the current practice of digital fabrication.

Furniture

David Yamnitsky took the How to Make (almost) Anything class in 2013. One of my favorite assignments in the class is to have the students "make something big." This assignment is given when I teach the students how to use the largest tool, an automated milling machine that can handle a four-by-eight-foot sheet of plywood given to each student. Yamnitsky's girlfriend wanted to work at a standing desk but couldn't find a suitable one

Fab lab furniture from David (*top left*), Ohad (*top right*), Queen (*bottom center and right*), and Amanda (*bottom left*). *David Yamnitsky, Ohad Meyuhas, Grace Copplestone, Amanda Ghassaei*

to buy, so he made a custom one for her for that assignment. To make the desk, he designed and cut out a kit of parts—similar to the flat-pack furniture bought from a big-box store. But rather than being mass-produced, every piece can be different. So many other people saw his project and requested variants of it that it became a Kickstarter project funded many times over. One of my students, Amanda Ghassaei, had the same difficulty finding a standing desk that she liked. Unlike David's desk, the one she wanted would be filled with storage underneath for her work in progress. After designing the desk, she was able to plot it out and assemble it in an afternoon.

A fab lab opened in Rwanda in 2016. In Kigali, furniture is made and bought in the Gakinjiro district, a messy place filled with groups of guys using manual tools. A young woman named Queen (Reine Imanishimwe) discovered that she could bypass this furniture district and make her designs in the fab lab as David and Amanda had done. She proceeded to populate the fab lab with its own custom furniture, which she made with wood recycled from a kitchen ceiling.

After attending the Fab Academy in 2012, Ohad Meyuhas then opened a fab lab in Holon, south of Tel Aviv, in 2013. Holon is a poor neighborhood with a large immigrant population and a high crime rate. Next door to the fab lab is a community center that was meant to be a safe haven for local kids but was a neglected, gloomy space. In 2014, he used the fab lab to do an extreme makeover, in which the kids got to design functional furniture for the space and then, in an intense week, produce everything. The result was not only a lively space that they now take pride in but also the kids' inspiration to grow up to become designers and makers.

Housing

The Shelter 2.0 project was started by two carpenters, Robert Bridges and Bill Young, who were early adopters of digital fabrication. That experience led Bill to work for ShopBot, maker of a popular large-format milling machine used in fab labs. Inspired by stories of the need for rapidly erected shelters for disaster response, they were struck by the possibility of producing it on-site, on demand, and customized to individual needs. They came up with an open-source design, shared the files, and did a trial of sending flat-packed shelters to Haiti, where they were rapidly erected by the recipients. These shelters were much more substantial than the surrounding informal housing and were produced at a cost of just a few dollars per square foot.

When he was growing up in Harlem, Larry Sass's love of architecture led him to MIT, where he was one of the first faculty members working with CBA. He became interested in the potential that digital fabrication held for mass customization of housing, bringing custom construction from a select few to everyone. After seeing people in New Orleans living in Federal Emergency Management Agency trailers a year after Hurricane Katrina, he began meeting with local homeowners. The result was his development of a digitally fabricated house for New Orleans. This was in the shotgun style, which has been suggested to reflect African and Haitian influences in American house design, with ornamentation based on existing shotgun houses in New Orleans. After testing the design with a laser-cut model, he (with help from Bill Young) then scaled up and cut out the parts to make it full size. This version, which needed only a rubber mallet for its assembly, was first erected and then exhibited in the courtyard of the Museum of Modern Art in New York City in 2008.

Digitally fabricated houses (*top to bottom*): Shelter 2.0, Digitally Fabricated House for New Orleans, Fab Lab House. *Bill Young, Larry Sass, Fab Lab Barcelona/IAAC*

Vicente Guallart founded the Institute for Advanced Architecture of Catalonia, site of the first fab lab in Barcelona, and later became the city's chief architect. With Tomas Diez and Dani Ibañez leading a team from twenty-five countries, they created a fab lab house that won the People's Choice award in the Solar Decathlon Europe in 2010. Although several other projects have sought to make houses with giant 3D printers—an approach that requires an enormous capital investment and has had limited progress—this team instead cut out unique kits of parts that could be assembled on-site, to erect a complete solar house, including its contents.

Guallart's design optimized solar power production and natural ventilation as an integral part of the construction. The shape, variously described as a "peanut house," "cinnamon submarine," "forest zeppelin" or "whale belly," efficiently produced more energy than it consumed. Along with receiving orders to replicate the house, the team turned it into an adjustable family of house designs that could be varied from a modest cottage to a large villa.

Flight

Of all the types of robots made in fab labs, the most popular may be those that fly: drones. The interest in drones ranges from recreational to professional. Fab labs can make a drone's mechanical structure, the propellers, the power electronics to drive the motor, the control system to guide it, and the communication system to talk to it. By making these elements in the fab lab, makers can customize a drone for a mission rather than simply selecting a gadget from a catalog. Technically, the motors could also be made in the fab lab, but motors are mass-produced commodities that don't require this kind of customization.

Matt Norris is an aerospace engineer who started a fab lab in Tulsa. One project there brings in local schoolteachers, who learn how to make as well as fly drones. The teachers then bring this experience to their students, helping interest them in learning about all the technologies in a drone rather than using it as a black box.

Chirag Rangholia studied architecture in India before attending the Fab Academy in Barcelona in 2014. He developed a drone for his final project, integrating it with a programmable camera mount that Aldo Sollazzo had developed for his final project. The two men have since turned their project into an organization, Networking Environmental Robotics (NERO). Rather than selling drones, NERO provides the data its drones can collect. For Barcelona's Department of Urban Development, NERO provides NDVI (normalized difference vegetation index) measurements to help the city map and manage its plants. In Costa Rica, the organization produces agricultural maps to help combat the proliferation of pests. Doing it this way is both cheaper and more easily customized than satellite imagery.

Daniele Ingrassia was working as a researcher at an Italian IT company when first he visited a fab lab, in Torino. He was so inspired by what he saw that it led him to enroll in the Fab Academy in 2015, at the Opendot lab in Milan. His final project was making another drone, this one incorporating a navigation system that allows it to reach destinations while avoiding obstacles. He enjoyed the experience so much he quit his

Family of fab lab drones. *Daniele Ingrassia*

day job and was subsequently hired through the Fab Economy website (a commerce platform for fab labs) by the Rhine-Waal University of Applied Sciences. Along with running their fab lab and serving as a Fab Academy instructor, he's continuing to investigate drones; he is currently creating a family of various sized drones that can be made in a fab lab and is integrating higher-level reasoning into their control systems.

Jonathan Ledgard was the longtime Africa correspondent of the *Economist*. Now based at the École Polytechnique Fédérale de Lausanne, he's working with the British architect Lord Norman Foster on developing droneports in Africa. Foster+Partners has designed some of the world's largest airports; these droneports will be the smallest. Only a third of Africans live within two kilometers of an all-season road. Rather than pave the continent, the goal is to provide aerial connectivity, concentrating initially on high-value payloads, including medicine and spare parts for critical machines like pumps. Ledgard and Foster are starting in Rwanda, a country that combines a rough geographical landscape with good government. When I was in Rwanda to deploy the first fab lab there, Ledgard approached me at a meeting of African leaders and asked about putting fab labs in these droneports to repair the drones. I explained that they could go a step further and make the drones in the droneports. These aren't simple hovering quadcopters; for long-range efficiency, they're fixed-wing aircraft, requiring the production of three-dimensional tooling forms that are used to lay up fiber resin composites to make high-performance structures. My student Grace Copplestone returned to Rwanda to work with the fab lab on an initial demonstration of the processes to produce a drone. That experience inspired the creation of a local drone development group in Rwanda.

The biggest drones of all are made by a former student, Yael Maguire, who heads Facebook's connectivity lab. With the wingspan of an airliner, these aircraft are for long-duration, high-altitude, solar-powered flight to provide global Internet access. Today, these drones require a manufacturing investment along the lines of what's needed to make an airliner. But as the research on discretely assembling large-scale composites progresses (see Chapter 5), even these will come within the reach of a future fab lab.

Communications

In 2002, I was approached by the Norwegian telecommunications company Telenor about participating in a smart home project in Oslo. I explained that there was such a proliferation of those projects then that few people I knew would be interested. I then asked if Telenor was doing anything different. After some awkward shuffling, the person mentioned that the company was working with an eccentric herder who was putting cell phones on sheep and reindeer in the far north of Norway. That herder was Haakon Karlsen, and he was doing this to track his flocks. Traditional Sami herding is nomadic, following the animals; with changing patterns of land use, the herders increasingly stay fixed but wanted to remain connected remotely. Haakon's project led to a collaboration to set up a fab lab, initially focusing on making the antennas and radios to extend wireless networks beyond the reach of cell towers. The project was then picked up at a fab lab in Pretoria, South Africa, to extend Internet access to where it was not easily available from the heavily regulated telecom provider. My student Amy Sun, with Keith Berkoben and Smári McCarthy (now representing the Pirate Party in Iceland's parliament), next brought the project to a fab lab in Jalalabad, Afghanistan, where there was no Internet access at all. The Jalalabad project grew to encompass a trial connecting forty-five sites, with the longest link between them being six kilometers. This project, which had come to be called FabFi, was then brought to a fab lab in Kenya, where it was deployed on a commercial basis to more than fifty sites. Instead of a communications infrastructure set up and sold by a large commercial carrier, this was community wireless operating regionally.

Data networks and communication terminals made in fab labs. *Amy Sun, Keith Berkoben, Smári McCarthy; David Cranor, Max Lobovsky*

To connect to one of these networks, for about a hundred dollars a fab lab can assemble a single-board computer with a small screen and keyboard to make a simple custom laptop. But you can provide access for much less than that; Max Lobovsky and David Cranor (who founded the 3D printer company Formlabs) took the How to Make (almost) Anything class in 2009. They continued working on what was called a thinner client. A *thin client* is an old idea in computing; it is a simple computer that has no local storage and keeps everything on remote servers. The modern version of that is the popular Chromebook, which can run just a web browser. Lobovsky and Cranor designed a minimal thin client computer that could be made in a fab lab with a bill of materials that cost just a few dollars in parts. It could generate video for a screen, could interface with a keyboard, and could communicate with the Internet. This computer could be mass-produced, but because they were made locally, each one could vary in its specifications, such as optionally adding local storage, varying the graphics resolution, incorporating assistive technologies, or including a radio for portability. And by making it locally, its makers gained the skills as well as the economic activity of a local investment.

Machines

The technical goal of a fab lab is to be able to make another fab lab. A small precision tabletop milling machine is one of the most popular tools in a fab lab today (and the one that I use the most). In 2011, after taking a class that I periodically teach on machine building, How to

Machines made in fab labs for producing parts and food. *Nadya Peek, Guillaume Teyssie*

Make Something That Makes (almost) Anything, Nadya Peek and Jonathan Ward developed a version of one of these machines that can be made in a fab lab: the MTM Snap (*MTM* for "machines that make"). A larger milling machine is used to cut the parts out of high-density polyethylene, a high-strength, recyclable plastic. Rather than being held together with fasteners, these are designed with flexible couplings built in so that the whole machine snaps together. It's driven by custom motors that are mass-produced, with the lead screws used to drive the machine built in. And the machine can cut out its own circuit boards for the motor controllers. Nadya and Jonathan's design inspired commercial products from Handibot and the Other Machine Co., and the personal fabrication roadmap that we'll return to in Chapter 5.

The same skills show up in making machines that make many other kinds of things, like food. Food production doesn't require breakthroughs in molecular biology; improvements can come simply from better integration of how it has been done for millennia. Guillaume Teyssie, working at a fab lab in Valldaura (in the hills outside Barcelona) that focuses on sustainable production, developed a system for aquaponics as his project for the Fab Academy in 2016.

Aquaponics is based on a symbiotic relationship between the aquaculture of fish and the hydroponic growth of plants, with the fish fertilizing the plants, and the plants filtering the water for the fish. This kind of system can be much more efficient than sticking seeds in a field. With aquaponics, you can precisely control the agricultural inputs and can expand your system vertically rather than horizontally for dense urban use. But it requires housings for the plants and fish, plumbing to circulate the water between them, and sensors, heaters, and lights in a control system.

All this aquaponics equipment is available commercially, but by making it in the fab lab, Teyssie could customize it to what he wanted to grow, along with saving the overhead of buying it from a remote vendor. His system has developed into the Aquapioneers project. With this system so far, he and his colleagues have grown their own lettuce, celery, beans, broccoli, cauliflower, strawberries, mint, basil, and coriander in the fab lab.

IMPLICATION

After someone asks what fab labs can produce, the next question is usually who pays for it. Although that's typically posed as an obstacle to the wider adoption of fab labs, lurking in this question is an even greater opportunity in the economic impact that fab labs can have.

The first fab labs that CBA deployed followed the same financial model that Andrew Carnegie used with his libraries. We donated the equipment and installed it; the sites had to commit to providing the space and the people to run it. Along with stretching our initial investment, this

Collaborating communities in the Holon fab lab. *Ohad Meyuhas*

commitment proved to be essential in establishing a local sense of ownership. Since then, fab labs have been funded by a variety of sources—public, private, philanthropic, for-profit—but they generally all have to take over their running costs.

The obvious way to assume your running costs is to sell things made in the lab. A factor of five is the typical ratio between how much a commercial product sells for and how much its materials cost, given the overhead of all the steps, from design to manufacturing to assembly to shipping to sales. You can cut out much of those steps' costs when they're all done at the same place and time locally. Given the opportunity for both cost reduction and customization, we paired successful entrepreneurs with fab lab inventors to build businesses around their prototypes. Our efforts were pretty consistently a failure. The entrepreneurs felt that the inventors weren't following their instructions, and the inventors felt that the entrepreneurs weren't telling them anything useful. It was hard to make enough of anything within a lab to sustain a business, and although some labs did manage to send their designs off for mass production, this step defeated the goal of bringing back the economic activity.

The problem was that this approach was a version of fighting the last war. Recall that it took Google a number of years to settle on the model of giving away its search capabilities and selling the benefits of searching (through targeted advertising). In the same way, the most interesting fab lab business models don't sell things that are made; they sell the benefits of making them. For example, Blair Evans's lab in Detroit works with at-risk youth, including pregnant teens and kids in the juvenile justice system. An important part of his funding comes from showing that the investment to engage these kids in the fab lab delivers better life outcomes than does the spending on existing social services. What he's producing is the transformation of individuals.

These models require as much invention as the technology does but are not (yet) taught in the canon of business schools. Successful fab labs have settled on an all-of-the-above approach to funding, with a mix of open time for the community, restricted time for members, teaching classes, supporting businesses, and producing infrastructure. This approach has been a feature rather than a bug, because any one of these sources of funding in isolation limits the lab; there are synergies across them. And common to all of these funding models is the need for the lab to be part of a larger network; a single lab doesn't have the ability to offer this full range of activities.

Funding for fab labs could be separated into for-profit versus nonprofit models, but at its heart is an even bigger idea that could be called post-profit. The loss of jobs to globalization and automation and the damages done by an economic race to the bottom underpin social and political up-heavals around the world. Yet implicit in all sides of the debate over com-peting financial and social models is an assumption about the nature of work. For many people, it means traveling away from home to get to a job they'd rather not be doing, producing a product designed by someone they don't know for someone they'll never see, to make money to buy what they need and want. What if you could skip all that and just make it for yourself?

There's a precedent for what appear to be economic facts of life turn-ing out to rest on implicit technological assumptions. *Peak oil*, the long-projected moment when oil production declines, was seen as a looming crisis. But not projected was that peak oil appears to be happening even sooner than expected because of the improving economics of renewables rather than the absence of oil. What if we're now approaching peak money, when the ability of a country to meet the needs of its population is no lon-ger measured by the output of its businesses?

This vision of consumers becoming creators resulted in the appoint-ment of Vicente Guallart, founder of Barcelona's first fab lab at the Insti-tute for Advanced Architecture of Catalonia, as the city's chief architect in 2011. The economic disruption was particularly acute there, with a youth unemployment rate exceeding 50 percent—a whole generation had no realistic prospect of finding work and living independently. Even so, ships continue to arrive in the harbor carrying products made far away, and trash trucks leave the city with waste destined for dumps. Guallart describes the current city as being based on converting products to trash. His goal is for electronic bits to come and go freely but for the atoms to stay in the city. He wants the city to make the transition from products-in/trash-out to data-in/data-out. To accomplish this, Barcelona is setting up fab labs around the city as part of the urban infrastructure. In the same way that a city is expected to provide clean water and electricity, they're providing the means to make. This approach wasn't uniformly embraced. In a poor immigrant district at the edge of the city, Ciutat Meridiana, there was a protest in which the community occupied the proposed site and de-manded a food bank instead. The disagreement was resolved as the pro-testers came to understand that they could use a community fab lab to help grow food, to make toys for their children instead of buying them, or to start businesses instead of searching for work.

Barcelona hosted FAB10, the tenth gathering of the fab lab network, in 2014. At that event the then mayor, Xavier Trias, pushed a button to start a forty-year countdown to urban self-sufficiency. Based on the digital fabrication research roadmap, they began by going from one fab lab in the city to one in each of the ten districts. The effort ends when the city can produce what it consumes. This goal could be viewed as a return to the city-state, not out of Catalan separatism, but because neither Madrid nor Brussels has much on offer to help with Barcelona's problems, so the people of Barcelona are finding their own solutions. The difference this time around is that they're not doing it in isolation but rather as part of a global network.

In 2013, Rep. Bill Foster (D-IL) first submitted legislation to do something similar to the Barcelona program in the United States, on a national scale. Foster has a remarkable background, first starting a business that pioneered computer-controlled stage lighting, then as a physicist leading the design of major components of the giant Fermilab particle accelerator, and then as a rare scientist in Congress. His National Fab Lab Network Act is written not as an appropriation but as the chartering of a national network of local labs in the national interest. Instead of an isolated resource inaccessible to most people, a national fab lab initiative would complement the existing labs by bringing the lab to the people. In an era of extreme political polarization, Foster's cosponsors come from urban and rural districts, representing both Republicans and Democrats. It is a rare issue today that can cross the aisle. It didn't come to a vote in that session (the US Congress was operating at something less than peak efficiency). But the legislation is being resubmitted and has already inspired private commitments that are aligned with its goals, starting with a ten-million-dollar pledge from Chevron to the Fab Foundation to set up fab labs in communities where Chevron works.

CBA has a mobile fab lab that we brought to the White House for the first Maker Faire there in 2014. It was parked right outside the Oval Office, where even people with White House badges aren't allowed to go. Security personnel were alarmed to see high-power lasers and powerful machine tools right there, but as an old community activist, President Obama loved it. The ostensible message was celebrating the maker movement. But there was a deeper subtext to the photo op: the event was highlighting that the new jobs are unlikely to return to the old factories. Just as personal computers were viewed as toys by the minicomputer industry before being destroyed by them, personal fabrication is likely to disrupt impersonal fabrication.

In 2015, FAB11, the next meeting in the annual gathering of fab labs, brought participants from seventy-eight countries to MIT's campus. At that event, leaders from Boston, Somerville, and Cambridge joined Barcelona's commitment. They didn't each get their own clock; they're all on the same countdown that Barcelona started. Many other cities have since joined them in what's come to be called the Fab City Pledge, now run by Tomas Diez.

Smart cities, namely, cities in which everything is instrumented and connected to be responsive in real time, have been a popular trend in urban planning. A fab city is the natural next step, crossing from digital to physical and able to sustainably produce and recycle what it consumes. This shift will not be a step change. Instead, it's a continuous transition or, more properly, an evolution by punctuated equilibria as new capabilities are introduced. Early steps include producing things like metropolitan wireless data and sensor networks, and furniture for civic spaces. Like fab labs, no one city has the skills to do all of this; they're being developed jointly by the participating cities, starting with quantifying all the poorly measured inputs and outputs to a city to track its progress.

The Fab City commitment is now inspiring commitments from countries. In 2014, when I was on a trip to Bhutan to plan a fab lab there, the opportunity was made concrete in a conversation I had with the thoughtful prime minister Tshering Tobgay about rice cookers. Bhutan is known for basing its economy around gross national happiness rather than gross national product. This doesn't mean that everyone is happy. It means that the Bhutanese are very serious about measuring the quality of how people live rather than how much they buy. This concern, unfortunately, doesn't extend to where things come from, which generally means trucks coming over the border from India. Rice cookers are central to the life of a household in Bhutan and are typically imported from Japan. Making the container for a rice cooker and adding a temperature sensor, a heating element, and a control system is an easy fab lab project. Along with eliminating the import of finished rice cookers, local fabrication allows each cooker to be customized to the size of the family and the design of the kitchen.

I had a similar conversation with Rwanda's minister of trade and industry, François Kanimba, in 2016 when I was deploying a fab lab there. He was one of the main architects of the economic reforms that allowed Rwanda to become one of the fastest-growing economies in sub-Saharan Africa. When we spoke, his greatest concern was Rwanda's large and growing trade deficit. The Rwandans were trying to tackle this with import substitution, which generally meant enticing multinationals to open factories in the country rather than import from factories outside the country. Distributing the means of production, rather than finished products, had not occurred to them until the opening of the first fab lab in Kigali. Just as Africa largely skipped landline telephones and went right to mobile, it can skip the historical stages of the industrial revolution and go directly to distributed production.

FAB12, the gathering of the fab lab network in 2016, was hosted by the city of Shenzhen. This is the heart of China's mass-manufacturing ecosystem, the engine for the loss of jobs in so many other places. We were there because, paradoxically, China is embracing fab labs and the maker movement, and these movements are embracing Shenzhen.

The Huaqiangbei district in Shenzhen is one of my favorite places on earth. It contains a massive market, where you can buy a single electronics component at one of the many small counters, or a bag of components from the counter, or a box from a closet down the hall, or truckload from the warehouse down the road, or a container load from the factory a few towns over. The market features what are called Shanzhai products, which I think of as the technological equivalent of rapid eye movement (REM) sleep. In the same way that your dreams mix apparently unrelated experiences into fantastical sequences, these products are technological mashups. The equivalent of sleep for their manufacturers is what they produce around large export orders. On my last trip to China, I bought what looks like an Apple watch, with two differences. One is that it cost $25 instead of $250. And the second is that it has a slot for the SIM card that Apple forgot to include—the watch is not a peripheral to a phone; it *is* the phone. What looks like intellectual property theft to the rest of the world is locally viewed as a flourishing open-source engineering community. Like classical composers, the producers shamelessly borrow themes and variations from one another.

If in the future Barcelona or Bhutan are going to produce what it consumes, their populations will no longer need Shenzhen to do it for them. The leadership in Shenzhen can see that the era of mass-producing goods, in which they've been so successful, is drawing to a close. However, for many years hence, the building blocks for this vision will be difficult to produce locally. These items aren't the finished products or even the machines that make them. They are the components of those machines. Shenzhen is making a giant pivot to mass-produce things like the motors with integrated lead screws for driving machines that Nadya and Jonathan used, or embedded radios for connecting devices to wireless networks with built-in Internet protocols. Because these elements require significant capital investments to produce, have economies of scale, and don't require customization, it makes sense to manufacture them in volume. In doing this, Shenzhen has been agile in merging multiple functions into a single item, to simplify the subsequent engineering effort.

There is a parallel with what happened to the computer industry after the arrival of personal computing, when the whole minicomputer industry disappeared. The commercial parallel to minicomputers for the third digital revolution is midsized manufacturers. But mainframes have become more important than ever, in the guise of the giant data centers that host cloud computing. The front end of computing for most people is smartphones, tablets, or PCs, but they rely on a back end in the cloud. Fab labs don't replace mass manufacturing; they extend it.

Think about what's happened to software or music. At one time, software was written by giants like Microsoft and IBM. Then came open-source software and the possibility that anyone could contribute. Today we have platforms where an app can be written and distributed for one person, ten people, a hundred, a thousand, or a million. The large proprietary software packages still exist but have arguably become the least interesting part of software development, because they have to target common denominators. What has opened up are tiers of software markets that were not previously viable. Likewise, music used to be sold by the record labels. Then came file sharing and a brief spike during which no one paid anyone for anything. Today there are platforms for buying and selling music tracks, with, again, markets of one, ten, a hundred, a thousand, or a million buyers. In the first case, a string of data becomes a program. In the second case, a string of data becomes a sound. Now, a string of data can become a thing. Mass manufacturing will continue to make products whenever people's needs are identical, and between mass production and do-it-yourself lies a whole hierarchy of new scales of manufacturing that are opening up and that were previously not commercially viable.

In the first two digital revolutions, there was a hope that a long tail of smaller content creators would power and be empowered by a new economy. Exactly the opposite has happened. The bulk of the money has been made by a small number of what have become giant companies that own the platforms—companies like Google, Facebook, Apple, and Amazon. Several groups have attempted to carve out a similar share in the third digital revolution, cornering the market for sharing free and paid designs of things to make. Some of these efforts are popular, but none have remotely reached the scale of the music and app sites. Perhaps the effort is just premature. A fab lab today needs all its capabilities to make a range of finished products; there are only so many small useful pieces of plastic that need to be made on an entry-level 3D printer. But another reason may be that making things is more like cooking than choosing entertainment. Although there are sites to share recipes, the commercial activity of cooking is centered around selling groceries and appliances.

Look at what happened with 2D printers. MIT spun off the Digital Equipment Corporation (DEC), which spawned the minicomputer industry. Then DEC failed and was bought by Compaq, which failed and was bought by HP. One reason HP survived was the economics of inkjet printing. HP's inkjet division is in Corvallis, Oregon, because the printer people had to hide from their management in Palo Alto, California. A group of engineers thought that they could make an inexpensive printer that would produce beautiful pages, but it would be slow. At the time, the printer division had a hierarchy of how many pages per second a printer could produce; the feedback from management was that a slow printer was a terrible idea. So the engineers moved to Corvallis, where the lowly calculator division was, and used this location as a cover to develop inkjet printers. The point now, of course, is that you don't need a high-volume printer on your desk, because every page you print there is different. There might then be a laser printer down the hall for a workgroup, a line printer in the basement for an organization, and a roll-to-roll printer in a warehouse for a city. But all the little inkjet printers producing a page at a time add up to the output of the giant roll-to-roll printer. Neither is better; what matters is how much what's being produced varies. This hierarchy in the migration of publishing onto the desktop is analogous to the coming migration of fabrication.

The scaling of capacity can go in both directions. In 2016, the White House hosted a follow-up event on the maker-to-manufacturing transition, on how to go from a prototype to a product. At that event, my student Nadya Peek dared the assembled group to think about exactly the opposite direction, from manufacturing to maker. The technology emerging for the transition in fab labs from rapid prototyping to the rapid prototyping *of* rapid-prototyping machines is a kind of automation, but rapid automation that can change to reflect needs rather than represent a large fixed-capital investment. Fears that automation will displace workers have assumed a rigid separation between *us* (the workers) and *them* (the robot owners). But the lesson of the third digital revolution is that *them* is us—ownership of manufacturing can become as widely distributed as ownership of computing did.

The most important kind of scaling for economic impact is not production capacity, but ideas. A 2015 report adding up the output of businesses spun off from MIT found that it was $1.9 trillion in annual revenue, which fell somewhere between the output of Russia, the world's ninth-largest economy, and India, the tenth. How can a few thousand people at a time match the productivity of a billion? There are two secrets to this.

The first is that MIT isn't an isolated technology park trying to make money. It is embedded in an ecosystem in Kendall Square—an environment that mixes long-term research, short-term development, small start-ups, and large corporations, along with cafés and clubs and parks. Little attention is paid to defining which activity belongs where.

And the second is that I think of MIT's core competence as being a safe place for strange people. Many of my colleagues would be considered dysfunctional in polite society. But by definition, to invent, you need to question assumptions, and that's not narrowly confined to a small slice of life. Much of this business creation doesn't start with dreams of riches. One of the drivers is a vision for change in the world, like, say, the need for flying cars. And an even greater driver is finding places to work where you fit in. A group of my students who worked on early quantum computing started a company (ThingMagic) to continue working together; their company ended up developing the reference platform for reading radio-frequency identification (RFID) product tags at store checkouts. They didn't start with a vision of conquering retail technology but stumbled across a match between a need and their skills.

The opportunity for fab cities, towns, and villages is to do the same on a global scale. The same secrets that make MIT work in the zip code 02139 are behind the spread of fab labs everywhere else. They're functioning as nodes in an ecosystem, defined not geographically but intellectually, where inventive people who don't fit in rigid school or business hierarchies belong. This kind of social infrastructure might appear to be intangible, but as the example of MIT spin-offs shows, it's even more important than the physical infrastructure. Countless regional economic development projects are spending vast sums to try to become the next Silicon Valley. But in the third digital revolution, the next economic engine isn't a place; it's a network of places linked by digital communications, computation, and fabrication.

ORGANIZATION

In 2005, I found myself on a boat with a broken motor, drifting ever closer to a glacier, off the coast of Svalbard, the last stop before the North Pole. I was there with Haakon Karlsen, the aforementioned herder who runs the fab lab in Lyngen at the top of Norway. We were visiting to look at the needs and opportunities in this most remote part of the country. The motor was eventually brought back to life, but those hours spent with an approaching glacier looming over us captured how it felt to try to keep up with the ever-increasing requests for access to a fab lab and people's offers to devote themselves to the movement. We realized then that we needed to build organizational capacity beyond what my lab or his farm could provide.

The problem was a classic commons issue. It wasn't hard to find funding for the parts of a fab lab that you could see. What was hard to fund was the infrastructure that made these parts possible—the infrastructure that you couldn't see. This included the development of the technology that went into a fab lab, the management of the global supply chain to source everything, maintaining the teams to deploy a lab, and the computing that supported them. All these infrastructure needs initially came out of my research funding, which was not a scalable model.

We had tried to partner with philanthropic foundations, but they lacked the technical skills to do this. People running investment funds did understand speculative technical risk, but they weren't interested in supporting commons. And research funders didn't support field organizations. As a last resort, we concluded that we had to create our own organization.

It took five tries to get it right. The first four attempts tried to raise funds to cover these common costs because it was a worthy thing to do. Those attempts didn't go anywhere. Not only were funders uninterested, but the decentralized fab labs didn't embrace the creation of a centralized entity. Instead of trying to sell the benefits of fab labs, the fifth attempt simply sold the services that everyone needed. That's how the Fab Foundation was started.

There's a common misconception that open-source projects are flat organizations. The large successful ones are based around a hierarchy with a benevolent dictator, such as Linus Torvalds with Linux, and Mitchell Baker with Mozilla. This strong leadership enables everyone else to contribute. Since people aren't compelled to follow this leadership, the authority comes from the soft power of motivating them to participate, rather than command and control.

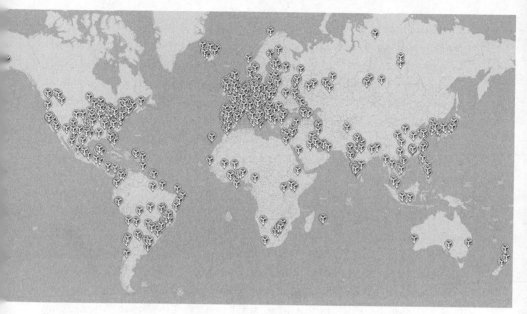

Fab lab locations in 2016. *Data: Fablabs.io. Map: © OpenStreetMap contributors at openstreetmap.org*

The most notable example of soft-power technological leadership is the Internet Architecture Board (IAB), which provides technical direction for the Internet. This is a committee of thirteen people who no one outside the networking community has heard of. The way they work is apparent in the title of the standards that they manage: requests for comment (RFCs). No one need follow these standards, but because of the benefits of being part of the Internet, almost everyone using a computer does.

The IAB is a committee of the Internet Engineering Task Force and an advisory body of the Internet Society. These sprang from the Internet Configuration Control Board, which was created in 1979 to advise what was then still a government research project. These organizations have lived on long past their modest origins, through the decades of exponential growth of the Internet.

The growth of the fab lab network is now at the same stage of spawning organizational scaffolding. The equivalent to RFCs includes a fab charter articulating the rights and responsibilities of a fab lab, the inventory of what goes into a fab lab, the curriculum at the Fab Academy, and a listing of where the fab labs are (at the fablabs.io portal). These elements all evolve with input from many people; any one decision may not be best for a particular purpose, but the collective benefit of having these common standards outweighs the cost of being optimal but fragmented. As the organizational capacity of the fab lab network has grown, it's being tapped by partner programs that want to work with fab labs collaboratively in a way they can't do individually. In one complementary approach, companies invest in a kind of have-your-cake-and-eat-it corporate social responsibility, supporting fab labs in communities where the companies are and, in so doing, identifying and training promising technical talent. In another approach, fab labs work with aid agencies to set up groups of fab labs in targeted parts of the world.

Patrick Colgan ran a European Union body to support post-Troubles reconciliation in Northern Ireland. The group's search for alternative ways to invest in the reconciliation led to the launch of fab labs at the Nerve Centre in Derry and at the Ashton Centre in Belfast. Both these centers are adjacent to the euphemistically named peace walls, which are really segregation barriers. Kids now come to work together in the fab labs from both sides of the wall, transcending their historical divisions. After seeing the impact that the labs were having, there is now interest in expanding this program to the rest of Ireland (North and South).

Engineering colleges in India can be notorious for divorcing formal study from building hands-on skills. In the state of Kerala, CBA set up fab labs initially in Kochi and Trivandrum. Now Kerala Startup Mission and Kerala Technological University are working with the Fab Foundation to establish the first 20 of a projected 150 fab labs based in engineering colleges. To provide the capacity for this, Fab Academy graduates are taking over from the initial MIT students to do the installation and training.

The first fab lab in Egypt was started in Cairo by Dina El-Zanfaly and a couple of Egyptian co-founders in 2011, after she took the How to Make (almost) Anything class as a graduate student at MIT. During the post-revolution riots in Egypt, she telephoned the lab to make sure that everyone was OK. Her counterparts in the lab laughed and reported that it was one of their busiest days, because the bright, inventive youth who had no interest in sectarian conflict had taken the disruption as a day off to go work in the lab. Their lab has since worked with the Egyptian Ministry of Education, USAID, and the Fab Foundation on establishing fab labs in STEM (science, technology, engineering, and math) high schools across Egypt, as well as with the Orange Foundation and other organizations on community fab labs, including the mobile Fab Lab on Wheels (FLOW). Last year, the lab organized Maker Faire Cairo, a one-day event that had ten thousand participants from the community.

Beno Juarez attended the Fab Academy in 2009 and then started Peru's first fab lab. This grew out of a collaboration between Barcelona's fab lab and Spain's aid ministry. Rather than sending money, Peruvian innovators were identified, trained, and then sent back with a fab lab. The lab upended the rigid educational hierarchy in Peru; when his university began to declare how the fab lab would be used, Juarez and his colleagues explained that it was theirs, not the school's, and unless it remained an accessible resource, they would move it elsewhere. He had grown up in the Amazon, where, he observes, there were three career choices: farmer, soldier, or terrorist. He chose "none-of-the-above" and is now leading a project to bring a floating fab lab up the Amazon River for sustainable production in indigenous communities.

We ran a pop-up fab lab at the United Nations for the launch of the Sustainable Development Goals (SDGs) initiative, which was the largest gathering of heads of state ever assembled. The goals came out of an inclusive process that followed what were effectively the Rich White Guys' Development Goals. The SDGs include objectives like access to health care, education, clean energy and water, and ending poverty and hunger. The assembled diplomats would look at us funnily, wondering why we were there. Then there would be an epiphany moment as they realized that most of the SDGs require the ability to go from bits to atoms, to locally make health-care sensors, water filters, and so forth.

We've also run pop-up fab labs at the World Economic Forum in Davos to show rather than tell the assembled political and business leaders about digital fabrication. At one of these, the then UN head of humanitarian relief encountered the lab, apparently annoyed that it was intruding on important discussions about providing infrastructure, educational opportunity, business incubation, and entertainment in refugee camps. Then you could see the light bulb go off as she realized that access to digital fabrication was a common denominator across all these issues—a commonality that the incumbents weren't considering.

This observation led to the launch of the Global Humanitarian Lab, run by David Ott, who came from the International Committee of the Red Cross, and Olivier Delarue, from the UN High Commissioner for Refugees. Their organizations are literally on the front lines of the most troubled parts of the planet. Ott and Delarue approached us about the possibility of shipping bits rather than atoms for rapid response, producing whatever is needed on demand locally, from prosthetics to shelters. When they were looking at setting up dedicated labs, they realized that they could do this as an overlay to the whole fab lab network in the same way that a range of services are now provided over the Internet. They could use the labs as a virtual platform for humanitarian relief.

In those kinds of gatherings of global leaders, I find that the politicians are generally rather glum. They're all struggling with intractable problems of unemployment, inequality, and immigration, along with the knock-on consequences of polarization and conflict. And the levers they've historically used to deal with these issues—adjusting things like monetary policy—aren't working.

On the other hand, the gatherings of the kinds of local leaders who, as I've described, are leading the third digital revolution on the ground are generally rather cheerful. They're having a real impact in their communities, there's lots of room for future growth, and, on top of it all, they're having fun.

With this divergence between top-down and bottom-up realities, there's a real sense that the leaders of the future aren't coming up through the organizations of the past. The current technology in a fab lab is intended to make itself obsolete, as I'll show in Chapter 5. Like the invention of the Internet on minicomputers, what's going to live on is not the current embodiment but the ecosystem that's evolving around the inventions. For the first thousand fab labs, it was interesting just to count their numbers, like following the appearance of the first websites. What matters now is not that they exist, but why. If anyone can make anything anywhere, how will we live, learn, work, and play?

N

ALAN & JOEL

CHAPTER 2

How to (almost) Make Anything

Anyone who has visited a thriving fab lab immediately picks up on the incredible energy, joy, and satisfaction that comes from harnessing the power of digital fabrication to design and make things. The early fab pioneers who Neil profiled, and others who we will introduce, are generating new models of highly distributed, personalized fabrication that may indeed transform how we live, learn, work, and play.

Neil highlights how people are increasingly able to make everything from food to furniture and from crafts to computers in community fab labs. These are early indications of how the third digital revolution will empower individuals and communities to become globally connected and locally self-sufficient, especially as the technology continues to exponentially improve. But Neil told only part of the story. Not every fab lab is thriving, and not everyone is thriving in a fab lab. Digital fabrication is a complex process involving multiple interdependent and evolving technologies and capabilities. While the potential exists for people to increasingly make what they consume, very few people are currently able to do so. Many of the raw materials used in today's fab labs are not renewable. There are large gaps between the potential for digital fabrication to transform society and the reality of it doing so.

Neil is a classic techno-optimist. His optimism is grounded in a deep understanding of the underlying science, his daily interactions with the pioneering early adopters, and the research roadmap. But he often greatly underestimates how social factors become significant rate limiters to the pace of technology development, as well as how the technology ultimately impacts (or doesn't impact) society. In this chapter, we present additional

perspectives on the current fab ecosystem, with the goal of providing social balance to the technology picture. While it is important to understand
the current capabilities of the technology and the success stories throughout the fab ecosystem, we also need to understand the challenges and
tensions that permeate the ecosystem. Building on what is working *and*
addressing what isn't are both essential to realizing the power and promise
of the third digital revolution.

The first two digital revolutions created great wealth and transformational changes, but they also left much of the planet behind. More than a
half century after the publication of Gordon Moore's paper, we still have
significant digital divides. Half the planet lacks access to digital technologies. In much of the world, a combination of income and wealth inequality,
technological unemployment, and digital echo chambers are deeply dividing society. Many people are struggling with an "always-on" life increasingly mediated by digital technologies.

The third digital revolution could help address these social challenges, or it could make them much worse. The pace at which these technologies emerge from the lab and their impact on society, for good or ill,
will not be driven by some invisible hand. Progress will be driven by the
decisions we make and the priorities we set, individually and collectively,
as the technologies are introduced into society. The best time to shape the
trajectory of accelerating technologies is early, when the research priorities are emerging, assumptions are being baked in, and the ecosystem of
supporting organizations and institutions is being formed. The time is now
for the third digital revolution.

In conducting research for this book, we have visited fab labs and
other maker spaces around the world, interviewed dozens of fab and maker
pioneers, and surveyed hundreds more. Across these communities, we
have observed a striking mix of exhilaration and frustration—a combination of deep optimism and serious concerns about the future of digital
fabrication. Neil highlighted the optimism and exhilaration in Chapter 1.
However, knowing what's possible with the technology doesn't guarantee
that this potential will be realized. We need to have a clear-eyed view of
these challenges to develop and advance the methods and mind-sets to
effectively address them.

It will not be easy. In early 2016, when the Defense Advanced Research Projects Agency (DARPA) announced its new initiative tackling
"next generation social science," Adam Russell, the program officer,
highlighted how the social sciences have been "inherently challenged

because of its subject matter: human beings, with all their complex variability and seeming unpredictability." He then continued, "Physicists have joked about how much more difficult their field would be if atoms or electrons had personalities, but that's exactly the situation faced by social scientists." Digital technologies detect and correct errors as they propagate. People also try to detect and correct errors, but the process is more complex and messy. Throughout this book, we look not only at the technology, but also at the less predictable human side of the third digital revolution.

The basic contours of human nature are not likely to change anytime soon, even as technological capability races ahead. The good news is that we will continue to dream, create, and shape our world, with individual and collective agency. The bad news is that this dreaming, creating, and shaping will not always tap into our better selves. Attention to the social dimensions of the third digital revolution requires examining underlying assumptions about human nature and the ability for individuals, organizations, and institutions to adapt to accelerating change. It may be easier to shape bits and atoms than people and society, but they are inextricably intertwined.

The title of this chapter is a play on the name of Neil's original How to Make (almost) Anything class. When projects failed, students joked that the class felt more like How to (almost) Make Anything. This sentiment often ripples through the broader fab community and speaks to the very real challenges that come with attempting to democratize manufacturing. In this chapter, we explore some of these threshold challenges, including issues around fab access, fab literacy, the cultivation of an enabling fab ecosystem, and the mitigation of risk as the technology propogates. In Chapter 4, we provide historical context for how social systems have been mostly reactive to new technologies and suggest proactive alternatives. We conclude in Chapter 6 with aspirational visions for fab futures that align social and technical systems to create a more self-sufficient, interconnected, and sustainable society along with specific guidance for addressing the challenges and transforming aspirational visions into reality.

FAB ACCESS

By 2017 there were about a thousand fab labs around the world, reaching a total of a few hundred thousand people. There are approximately seven billion people on the planet. Neil has shown how fab labs can transform people's lives, but there is a real risk that these benefits will be enjoyed by only a fortunate few. Throughout all our interviews, there was a common concern about a potentially dangerous *fab divide* if widespread access to digital fabrication technologies is not a priority and too many people are left behind. Thus, fab access is the first threshold challenge.

Every breakthrough in technology has created gaps in society. The first two digital revolutions created enormous wealth and massive change across the globe. But not everybody has benefited. Billions of people have no access to computers or the Internet, lacking the basic digital infrastructure necessary to participate in the third digital revolution. The growth of the Internet is both an exemplar for the third digital revolution, illustrating that exponential change is possible, and a reason for caution, given the pervasive digital divides.

In 1983, the TCP/IP (Transmission Control Protocol/Internet Protocol) made interoperability among the growing network of computers possible. By 1989, there were fifty thousand users of the Internet—mostly academic and military. This was the year that the ARPANET, funded by the US Department of Defense, was opened to the public. The historic pattern for access to the Internet is illustrated in the following figure, which indicates that the initial adoption was slow and then rose to the point that, today, half the world's population has some form of Internet access.

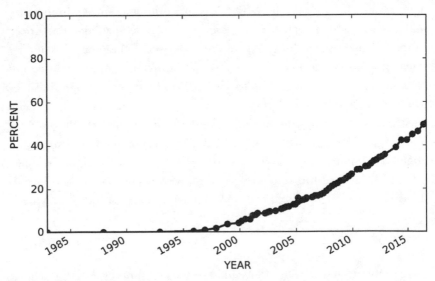

Users of the Internet as a percentage of the world's population. *Sources: Internet World Stats, International Data Corporation, Nua Ltd.*

The figure presents an impressive rate of growth. Note, however, that the rate of change is relatively linear, even though the growth in the technological performance was exponential, reflecting Moore's Law. As a result, the Internet's underlying technology has long been capable of serving 100 percent of the world's population. A combination of actions (or inactions) by individuals, organizations, and institutions accounts for the impressive reach of the Internet and the failure to reach even further. Today there are a growing number of "moon-shot" initiatives from governments, foundations, social entrepreneurs, and for-profit companies seeking innovative solutions to providing universal Internet access. What if this deep investment of human and financial capital, energy, and innovation had begun as early as 1983?

Moreover, Internet access is not a binary distinction, with simple haves and have-nots. Across the globe, there are substantial disparities in the quality and reliability of computing and Internet access, with billions of people having mobile-only access, inconsistent connectivity, or tiered access, with the faster tiers being out of financial reach. Try writing and sharing a document with rich media on a flip phone or a low-end smartphone; then imagine trying to navigate CAD (computer-aided design) and CAM (computer-aided manufacturing) workflows on that mobile phone.

Nor are Internet access and quality just developing-world problems. The Joan Ganz Cooney Center at Sesame Workshop, which does pioneering research on issues of digital inclusion, recently released a report on the digital divide in the United States. One-third of those below the poverty level in the United States rely on mobile-only Internet access, and many experience regular interruptions to their Internet service because of weak local infrastructure or an inability to pay the necessary monthly fees. Digital access is a continuum with real challenges for people at the lower end of the economic order. Given that nearly every job (and much of the rest of human activity) is now mediated by digital technologies, these digital divides have, not surprisingly, become key drivers of growing inequality in global wealth and income.

Within the digital fabrication world, affordable and reliable Internet access was cited as a clear threshold challenge by more than 175 fab leaders whom we surveyed in early 2017 (additional information on the survey methods and results are in the Resources section at the end of the book). Internet access was rated as very important by 80.4 percent of the respondents. At the same time, two-thirds (66.9 percent) reported that it was also very difficult. This gap between importance and difficulty represents a

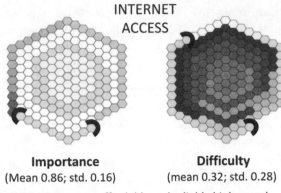

INTERNET
ACCESS

Importance
(Mean 0.86; std. 0.16)

Difficulty
(mean 0.32; std. 0.28)

*My Having Access to affordable and reliable high-speed
internet services (.54 gap between the two means).*

Importance and difficulty of access to the Internet. *WayMark Analytics*

threshold pain point constraining the exponential growth of digital fabrication. To understand these responses better, we use a data visualization approach that Joel developed with colleagues as a way to see points of alignment and misalignment among stakeholders.

In the above visualization, every respondent is assigned a small hexagon, which is color-coded to represent their views. These views are normally presented as red, yellow, and green, representing degrees of negative, neutral, or positive views. Because these illustrations are depicted in gray in the figure, the red is represented with darker shades of gray, yellow with middle shades of gray, and green with lighter shades. The small hexagons are tiled in a spiral fashion, beginning in the middle of each figure with those closest to the mean and then alternating above and below the mean until those farthest from the mean (literally, the outliers) are on the outside of the figure. The middle is the central tendency. The brackets are added to indicate "not applicable." In the case of Internet access, the contrast is clear—virtually all the respondents said that Internet access was important, and almost as many said it was hard to obtain.

The third digital revolution doesn't depend only on the technological infrastructure of the first two digital revolutions. It also introduces new technologies that could easily make the current disparities much worse. Since digital fabrication requires access to a continually evolving array of hardware, software, and consumable materials (the raw materials used in fabrication)—as well as space, computers, Internet connectivity, and qualified staffing—the challenges of fab access are even greater than the challenges of the first two digital revolutions.

In our survey of digital fabrication leaders, we asked about access to the needed software and hardware and the integration of software and hardware. Again, there were major gaps. Access to the equipment in a fab lab was seen as very important but hard to obtain. There was a similar pattern with software. The situation was only slightly better for the integration of software and equipment, as is indicated on the next page.

The responses on difficulty varied across software, equipment, and the interface between the two, with the most pain associated with software. When asked about difficulties with equipment, the responses on difficulty are bimodal, with over one-third indicating that equipment access was very difficult and another third indicating it was very easy. As a map of stakeholder interests, this split is particularly challenging. If we don't bring up the bottom third before enhancing things for the top third, we risk a growing fab divide.

As we noted in the introduction, when President Obama publicly declared that "high-speed Internet is a necessity, not a luxury" in 2015, it was a full half century after the publication of Moore's paper. If we wait another half century for a US president (or another world leader) to realize that fab access is necessity—just as electricity, water, and digital connectivity are—then there will almost certainly be creating a crippling fab divide. This, in turn, will exacerbate the existing digital divides, risking a future with even greater gaps in wealth and income, more technological unemployment, and further destabilization of society.

If, on the other hand, we proactively leverage digital fabrication technologies to increase personal and community self-sufficiency, the benefits will ripple through all aspects of society. This could change how we conceptualize the very concept of work and other aspects of everyday life. Working toward the goal of universal fab access will require a focused, sustained, and aligned effort by those pioneering both the technical and the social systems powering the third digital revolution.

While fab access is critical for everyone in society, it is particularly important for those who have been left behind in the first two digital revolutions. The most moving stories from our interviews came from fab labs in inner-city and rural settings, where youth have traditionally had limited access to tools for personal empowerment and self-sufficiency.

Mel King, the longtime Boston community activist who created the South End Technology Center fab lab with Neil, highlights the impact on urban youth: "We have a class here for folks who were formerly incarcerated or who have substance abuse issues. We have been doing this for

HARDWARE ACCESS

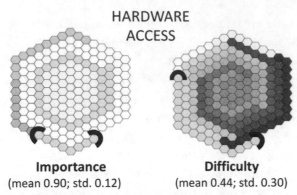

Importance
(mean 0.90; std. 0.12)

Difficulty
(mean 0.44; std. 0.30)

My having access to the needed equipment for digital fabrication (.56 gap between the two means)

SOFTWARE ACCESS

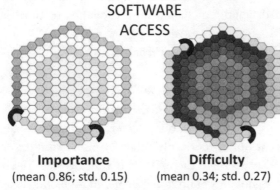

Importance
(mean 0.86; std. 0.15)

Difficulty
(mean 0.34; std. 0.27)

My having access to the needed software for digital fabrication (.52 gap between the two means).

INTEGRATING HARDWARE AND SOFTWARE

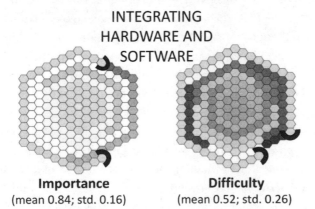

Importance
(mean 0.84; std. 0.16)

Difficulty
(mean 0.52; std. 0.26)

Integrating digital fabrication equipment with digital fabrication software. (.32 gap between the two means)

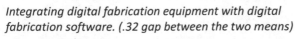

Importance and difficulty of access to hardware, software, and the integration of hardware and software. *WayMark Analytics*

a few years with the various agencies. Getting the folks into the fab lab brings back meaning for them. Seeing the expression on their faces when they are creating something, usually for a loved one, lets us know that we are on the right track. . . . I have seen lives transformed."

Another example comes from the Vigyan Ashram school in rural India, where the idea of fab labs was first piloted. Yogesh Kulkarni, who manages the lab, describes how it empowers local youth: "We work with youth who are fourteen to twenty years old, many of whom are school dropouts. At the fab lab, they develop a wide variety of skills in solving problems and learn the design process. We've had students go back to their villages and design hydroponic watering systems, which they then fabricate back here in the lab. Some of the students have been able to purchase single machines such as a laser cutter and set up a local service business. Others have been able to raise money and launch fab labs."

And the benefits won't be limited to youth and the previously disenfranchised. If the democratization of manufacturing can slow or rebalance the accelerating divide in income and wealth worldwide, it will benefit everyone. In *The Spirit Level: Why More Equal Societies Almost Always Do Better*, public health researchers Kate Pickett and Richard Wilkinson document how the healthiest and happiest societies have the narrowest gaps between the rich and poor. Digital fabrication technologies can be a key part of the solution in narrowing these gaps, but only if fab access is addressed in systematic and comprehensive ways. Chapter 6 further develops ways to address fab access and other threshold challenges.

FAB LITERACY

The second threshold challenge is ensuring that individuals know how to use fab technologies once they have access to them. We were decades into the first two digital revolutions before educators, employers, parents, and policy makers realized that digital literacy was essential to surviving and thriving in the digital age. The digital divide has been as much about literacy as access. Although there are various definitions for digital literacy, broadly speaking it involves developing the knowledge, skills, and mindsets needed to understand and use digital technologies to accomplish personal and professional goals as well as being a responsible digital citizen. Digital literacy includes everything from finding and consuming digital media to creating and communicating using digital technologies.

The majority of people with digital access consume digital content and use simple tools like Snapchat. Most people have a limited understanding of the digital platforms and technologies that shape their daily lives. Youth and young adults are spending an average of eight hours a day immersed in digital media, but they are not necessarily becoming digitally literate. Consider the massive youth and young-adult unemployment throughout the world; even in regions where there is deep mobile penetration, many technology jobs are going unfilled. This digital-creation gap, which is all about literacy, could easily be made much worse when creation goes from bits to atoms.

Digital fabrication is hard. It introduces a set of new competencies, including the navigation of continually evolving CAD and CAM software as well as additive and subtractive hardware, embedded computing, and an understanding of the biological and chemical properties of the materials used in fabrication. It also requires design thinking, creativity, collaboration, problem solving, and resiliency. These all require knowledge, skills, and mindsets that cross very different disciplines and domains and, as a result, are not currently well integrated. We define fab literacy as the social and technical competencies neccessary for leveraging digital fabrication technologies to accomplish personally and professionally meaningful goals, as well as a commitment to the responsible use of the technologies. We cannot build toward a more self-sufficient, interconnected, and sustainable society without widespread fab literacy.

Nadya Peek, a pioneer in personal fabrication at the Center for Bits and Atoms, argues that "the number of fab labs may be growing exponentially, but the number of people empowered by the machines is not growing exponentially." She attributes this gap to the significant complexity involved in nearly all aspects of digital fabrication: "The software is archaic and difficult to learn—people can learn to print a plastic thingamabob—but this doesn't mean they have full access to the means of production." In other words, access requires literacy.

Digital fabrication is essentially a new language. As literacy scholar James Paul Gee points out, "literacy is only possible if there is a grammar. Grammar enables communication and, simultaneously, limits your choices." Gee notes that grammar is to language as rules are to sports. A game of basketball or football is not viable unless all the participants agree to the rules. He also says that when you learn a language, you internalize perspectives without realizing it. Languages have affordances (enabling certain ways of thinking and actions) and limitations. Values and assumptions are built into grammar.

In his How to Make (almost) Anything class, Neil in effect introduced a language for digital fabrication. He initiated the creation of a grammar—a set of rules and, to a great extent, the culture. The Fab Academy and all the different emerging extensions (e.g., How to Grow (almost) Anything) could be considered dialects of this foundational language. Even social constructs such as the fab charter, which serves as a shared foundation for governance, operations, and growth of the fab lab movement, are part of this language. For a language to survive, it needs a foundational grammar but it must also be adaptable and extensible.

The maker movement, hacker spaces, TechShop, and other maker approaches all offer a variety of languages with diverse (and overlapping) grammar, rules, and cultures centering on how things get made. Each approach offers different strengths and weaknesses that appeal to different people. To cultivate fab literacy, we need to recognize that the fab community is developing new languages with foundational and emerging grammars, rules, and cultures. Building literacy at the level of the various actual languages used is key to socializing new entrants into the world of digital fabrication, as well as in facilitating dialogue and synergies across different fab, maker, hacker, and other communities.

Digital fabrication not only involves a new literacy, but also builds on older types of literacy. As Gee cautions, "literacy in digital fabrication is dependent on multiple other literacies such as reading, writing and a variety of digital literacies. These older literacies may be reconstituted in the context of digital fabrication, but there is still a dependency on them." To put the importance of traditional reading literacy in perspective, according to a 2013 study by the US Department of Education, thirty-two million adults in the United States can't read, and close to a quarter of the population (21 percent) reads below a fifth-grade level.

Another challenge related to fab literacy is the tools' ease of use. Many people throughout the ecosystem spoke of a strong desire for CAD/CAM work flows that were more intuitive and more accessible. But this desire was also matched by a deep concern that the tools not be so easy that they become black boxes, which allow people to make (almost) anything but don't show people how to make (almost) anything. A black box is really the antithesis of the fab ethos. Captain Picard most likely could not have fixed the *Star Trek* replicator if his Earl Grey tea no longer came out hot.

Jens Dyvik, founder of the Oslo, Norway, fab lab, describes this tension between ease of use and a black-box approach: "Neil's allusions to building the *Star Trek* replicator is a bit concerning. It encourages excessive consumption over empowering creation. . . . This is a terrible future.

You have your replicator at home, but so what. You have more stuff, but are you happy? Being able to easily make lots of stuff can be bad as well. That is a challenge that many people don't think about." When technologies are complex, there is a risk that only a small class of priests (the creators) will have all the agency and that the broader laity (the consumers) lack the knowledge of how to shape the technology. Right now, the fab ecosystem does have a high-priest class that is truly fluent in the language of digital fabrication. Consider the learning involved in just some of the topics introduced in Neil's How to Make (almost) Anything course:

1. Digital fabrication principles and practices: 1 week
2. CAD, CAM, and modeling: 1 week
3. Computer-controlled cutting: 1 week
4. Electronics design and production: 2 weeks
5. Computer-controlled machining: 1 week
6. Embedded programming: 1 week
7. 3D molding and casting: 1 week
8. Collaborative technical development and project management: 1 week
9. 3D scanning and printing: 1 week
10. Sensors, actuators, and displays: 2 weeks
11. Interface and application programming: 1 week
12. Embedded networking and communications: 1 week
13. Machine design: 2 weeks
14. Digital fabrication applications and implications: 1 week
15. Invention, intellectual property, and business models: 1 week
16. Digital fabrication project development: 2 weeks

Gains in fab literacy depend on gains in fab access. Like other forms of literacy, fab literacy requires enough time for individuals to progress from a limited working knowledge to proficiency to mastery to the ability to teach others. As with access, if we do not proactively lay the foundation for developing universal fab literacy, we will see the same widening of the existing divides that have been exacerbated by other literacy gaps, be they reading, writing, or digital literacy. If, however, we continue to develop new approaches to cultivating fab literacy, we are likely to see many benefits, ranging from specific workforce development skills to increasing personal capability.

Scott Simenson directs the Engineering Fab Lab and the Additive and Digital Manufacturing program at Century College in White Bear Lake, Minnesota, a two-year college preparing people for the workforce. Scott observes that "industry is adapting to digital processes. We need to prepare the workforce for the jobs that are coming." He shares the example of a group of five students invited to meet with representatives from Wilbert Plastic Services, a thermoforming manufacturing company that makes parts for the auto industry and other sectors. The company suspected that 3D printing could reduce the two to four weeks needed to produce tooling for the injection-molding equipment, but it did not have expertise in the technology. Wilbert asked if it could work with some of the students as consultants. Scott reports, "Over a series of three or four meetings, the students and the company officials identified jig fixtures for assembly lines that the students could prototype. In the first year of the program, students are already doing consulting at the leading edge of the field."

The gains in literacy go beyond workforce preparation to increased personal capability. Scott teaches a version of Neil's class, How to Make (almost) Anything, as part of a new two-year curriculum in additive and digital manufacturing. He originally planned for the capstone project to be for all the students to make their own 3D printer. In fact, four students are already doing this in the class. He jokes that he will now have to come up with a more challenging capstone project, but the deeper point is that these students are not just gaining literacy in using the equipment; they are developing both an understanding of the underlying technology and an ability to shape it.

ENABLING ECOSYSTEMS

Widespread fab access and literacy are foundational parts of a more self-sufficient, interconnected, and sustainable society. Access and literacy are not acquired in a vacuum; the context also matters. At present, the context for digital fabrication includes an array of interacting, independent, and countervailing elements. A threshold challenge is to shape the context into an enabling ecosystem—oriented toward greater self-sufficiency at individual and community levels.

We focus here on five elements of an enabling ecosystem. These include cultivating methods for more effective collaboration and knowledge sharing; ensuring widely distributed mentorship and leadership; creating open and robust marketplaces for fab products and services; attracting a diverse mix of public, private, and philanthropic financing; and designing effective forms of governance aligned with the emerging fab community values. Often, the larger context is seen as fixed—exogenous—but we are identifying aspects of this ecosystem that can (and must) be shaped if digital fabrication is to grow at exponential rates and improve lives and society.

Collaboration and Knowledge Sharing

A key element of an enabling ecosystem for digital fabrication is the ability for anyone, anywhere, to collaborate and share knowledge, advice, failures, and successes with anyone, anywhere. Since digital fabrication requires a wide range of expertise and since the technology is rapidly evolving, the ability for people and projects to be shared across the network is both necessary and a considerable challenge. With effective mechanisms for collaboration and knowledge sharing, people can accomplish collectively what they have difficulty doing individually: network effects become possible. Consistent with Metcalfe's Law, which Neil introduced in Chapter 1, collaboration begets more collaboration, knowledge sharing generates new knowledge, and digital fabrication is better able to grow at an accelerating rate and empower more people.

To achieve network effects, the digital fabrication ecosystem needs interoperability—social and technical. The architecture of the fab lab movement is designed, at least on the technical side, with this aim in mind. Fab labs seek a common footprint for digital fabrication hardware and software, facilitating collaboration and knowledge sharing within and across the global network of fab labs. Although there is a common hardware footprint among most fab labs, there is still a great deal of friction in the process. Digital fabrication is awash with a mishmash of open-source and proprietary CAD and CAM software. This software variety and variability results in a great deal of fragmentation and friction when it comes to collaboration across fab labs. There is further complexity in bridging across the various configurations and cultures of maker spaces, hacker spaces, TechShop workshops, and other similar maker-centric places and communities.

The challenge of interoperability and extensibility in the larger fab and maker ecosystem reveals a key tension: the desire for standardization and the resistance to hierarchical decision making. The process of developing standards that align the interests of interdependent but independent stakeholders is, of course, not new. Some standardization has been

COLLABORATION
AND NETWORKING
ACROSS SPACES

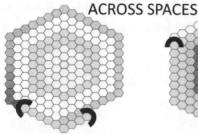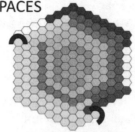

Importance	**Difficulty**
(mean 0.86; std. 0.16)	(mean 0.48; std. 0.26)

Collaboration and Social Networking among people in Fab Labs, Maker Spaces, Hacker Spaces and other locations involved with digital fabrication (.38 gap between the two means).

Importance and difficulty of collaboration across spaces. *WayMark Analytics*

essential to everything from consumer electronics to transportation to the Internet. History teaches a hard lesson here. As John Leslie King, former dean of the University of Michigan School of Information, sums it up, "standards are easy to establish early on, when no one cares about the standard, and almost impossible later on, when it is clear that the standard is important. Then everyone still wants a standard, but as long as everyone else adopts *their* standard."

To converge on standards for collaboration and knowledge sharing for digital fabrication, we need to address early on how diverse stakeholders will operate in an environment where no one person or organization is in charge. Progress will depend on effectively identifying interests, alignments, and misalignments in the current practices so that stakeholders (individuals and organizations) can advance their shared and separate interests. This process, which Joel and his research collaborative call *lateral alignment*, has been a challenge both within and across the different maker communities.

In our survey of digital fabrication leaders, we asked about their connections to different types of associated spaces, with the option to check all that apply. Even though the survey was distributed through the Fab Foundation email list, the responses indicate a great deal of overlap across the communities, with 81.1 percent of the respondents indicating a connection to a local fab lab, 42.3 percent to a maker space, 31.4 percent to a Maker Faire, and 17.1 percent to a hacker space. When we asked about collaboration and social networking across these spaces, as indicated on the prior page, three-quarters (77.5 percent) reported high importance, while over a third (36.9 percent) reported it was very difficult to do and more than half said it was at least somewhat difficult, indicating a key gap.

These communities may operate with considerable overlap and synergy, but there is also a subtle but pervasive tension around language, identity, and culture across fab labs, maker spaces, hacker spaces, TechShop facilities, and other emerging community resources for making. They are all growing in parallel, but there is tension around how much they can and should grow together. In the social sciences, we see that managing the tension between homophily (connecting with those like yourself) and diversity (appreciating those different from yourself) involves constant balance and iteration. Achieving the balance begins with an understanding of the roots of the similarities and differences.

Each of these communities has its own origin story. When we spoke with Dale Dougherty, founder and CEO of Maker Media, about the launching of *Make* magazine and Maker Faires, he used an analogy to the early gyms: "The first gyms were body-building workshops populated by serious, mostly male weightlifters. This was not welcoming for many people who wanted to work out but who weren't serious weight lifters." He thought about using the word *hack* or *hacker*, but felt that these terms would not be welcoming or friendly to many groups, especially schools. He also looked at fab labs (Neil was featured in first issue of *Make* magazine) but was concerned "that the cost and technical complexity of launching a fab lab was beyond the reach of most people." He wanted anyone, anywhere, to be able to start a maker space or participate in a Maker Faire with little cost and little difficulty. Ultimately, at his daughter's suggestion, he went with the simple word *make* because, as she put it, "everyone likes making things."

Interestingly, as both maker and fab communities have been spreading over the last decade, the two movements have been growing closer together in some ways. The cost of digital fabrication hardware and software, while still considerable, has steadily been reduced. And the technological sophistication of maker spaces has steadily increased. Dougherty jokes that he sees Neil "as the R&D arm of the maker movement." Neil embraces this description. In many ways, the fab lab movement is the cutting edge of the broader maker movement in terms of taking advantage of rapidly advancing digital fabrication technologies to more effectively make a broader range of things with both form and function at multiple scales.

Hacker spaces and TechShop have different origins and cultures. Hacker spaces have always been more decentralized, with a loose, informally distributed, and community-driven culture. The hacker space community has a volunteer-run website with a wiki, a blog, and other resources, but there is no founder or central organization, and each organization has different operating models grounded in how its local community or members define hacker culture. TechShop, which was launched in 2006, is "a membership-based, do-it-yourself workshop and prototyping studio that provides makers of all ages and skill levels affordable, open access to a wide range of tools, equipment, resources, and workspace."

TechShop features a wide variety of digital and nondigital tools and operates on a membership model. Many people are confused about the differences between these different maker-centric organizations and communities. In a 2013 article in *Make* magazine, Gui Cavalcanti attempts to explain the differences. Observing that many knowledgeable people don't distinguish between the terms *hackerspace* and *makerspace*: he counters, "I personally find that I need to differentiate between the two, because at this point the concepts and representations behind the words have diverged significantly for me."

Cavalcanti also addresses the TechShop and fab lab communities, both of which he sees as encompassed by the larger language of maker spaces: "In my mind, both TechShop and FabLab are makerspace franchises; they focus on creation from scratch, through multiple types of media. Ironically, both came into being before the term 'makerspace' was widely used, and as such their trademarked names have more staying power right now than the overarching term."

Of course, not everyone will agree that *maker* is the overarching term. Moreover, fab labs are neither a franchised business nor a trademark (although TechShop is both). Cavalcanti, however, does highlight how the shared identities across these communities are joined by real differences in approach, culture, and even core values. It is both natural and healthy for many maker-centric communities to emerge with varying capabilities, cultures, and operating practices. That said, an enabling ecosystem depends on communities' ability to collaborate and share knowledge—especially with respect to digital fabrication, where designs and fabrication can be digitally shared, adapted, and co-created. If diverse stakeholders can align around interoperable standards, then many more people and groups can benefit while maintaining their own identity and operating models.

Mentorship and Leadership

A key element of an enabling fab ecosystem is ensuring widely distributed leadership and mentorship across the community. Sherry Lassiter, who leads the Fab Foundation, points out the importance of people: "I used to think that growing the network was just about bringing down the cost of launching new fab labs. Cost is part of it, but it is much more the people. It is not just a matter of putting in new labs in a top-down way. To provide access to digital fabrication we need to build the human capability to go with the new labs and we need sustainable leadership of the labs that have been launched."

When people launch a new fab lab, the equipment will often be up and running in a matter of weeks or, at most, a few months. But getting the needed social system up and running is another story. It can take years to get the staffing, internal governance, sustainable financial models, and program design working right, not to mention establishing the necessary relations with key external stakeholders. So, right from the start, the social systems lag the technical ones in a fab lab.

Many of the early fab labs do have strong leadership, as they were founded by deeply committed social entrepreneurs who have worked through the diverse challenges that come with launching and running a fab lab. Less visible are the people who have attempted and failed to launch or sustain a fab lab. As the demand for community fabrication grows, a threshold challenge in the ecosystem will be the parallel need for growth in leadership capacity.

A related hurdle for the fab ecosystem is finding experienced mentors who have the time to support all the individuals who need help navigating their way around the software, hardware, and materials. Fab pioneers like Jens Dyvik and Nadya Peek, folks who have set up fab labs and trained local leadership, are much in demand. And yet, it's hard for them to balance their own research and passion projects with the growing number of requests for advice and mentorship. In fact, this was a common struggle among many of the fab pioneers we spoke with. Many fab labs don't have on-staff dedicated mentors who can help guide the wide variety of projects that come through their labs. This challenge for the larger fab ecosystem is likely to grow as the number of fab labs grow and as the technology enables fabrication to become more personal instead of community based.

In addition to the need for distributed on-site mentorship, there is also a need for a wide variety of resources providing just-in-time guidance and feedback for those who don't have access to a local mentor and who run into challenges while tackling a project. Overall, a large majority of fab leaders (87.3 percent) reported that access to mentors in their local labs was very important. At the same time, over a third (35.5 percent) said that this was very difficult to do, and nearly half (48.4) said that it was at least somewhat difficult to do.

Mentoring is not just about working with the software and hardware needed for digital fabrication. Another key knowledge or resource gap in the current fab ecosystem is guidance on the raw materials used in digital fabrication. This is an area that Alysia Garmulewicz, a professor at the Universidad de Santiago in Chile, is attempting to address. Her research is focused on developing a circular economy, where resources in fab labs can be sourced locally and continually reused, as opposed to a linear economy, where resources are extracted, used, and discarded. She points out, "We need to understand the materials that are around us so we can create a high-performance economy with natural polymers and other ingredients that are underused. This requires distributed information regarding the availability and usability of local materials." Currently, resources like this are very limited across the fab environment.

The challenge of accessing just-in-time knowledge and feedback can also be part of the design of the hardware and software. Nadya Peek describes the current complexity, the need for resources, and the opportunity for feedback to be built in the technology itself: "There are thousands of settings on a laser cutter for different materials—the frequency of the pulsing of the laser, the speed at which it moves, et cetera. I know this in my head and can answer questions when people ask, but there is no shared resource with this information." Peek adds that it is not just the details of the separate machines: "There are lots of complex components, and they all need to be standardized and interoperable, with ways to adjust to technology that is constantly evolving." She points out that "we need to develop the infrastructure so that knowledge and agency are implicit in the technology." It is critical, as Peek clearly highlights, that the hardware and software become more intuitive and provide better feedback to the users.

This idea of people and machines working effectively together is central to the fab experience and has been the subject of several recent books about the future of work. For example, MIT researchers Andrew McAfee and Eric Brynjolfsson, coauthors of *The Second Machine Age: Work, Progress, and Prosperity in a Time of Brilliant Technologies*, explore areas of the current and future economy, where the synergy between people and machines is greater than either individually. They, along with others who highlight the need to "race with the machines," often cite freestyle chess as an evocative example of technology complementing humans. In this type of chess competition, computers are paired up with humans to compete against supercomputers (which have beaten the world's top chess masters). In these competitions, teams of people and computers typically beat the top supercomputer or the top chess master playing individually. Designing fab technologies to work more effectively with the fab creators will be another key component of an enabling fab ecosystem.

Robust Marketplace Platforms

The fab ecosystem will also require open and robust marketplace platforms, which are the vehicles for the discovery, sharing, buying, and selling of digital fabrication designs, products, and services. A key lesson from the first two digital revolutions is that great wealth and influence accrue to those who control the leading marketplace platforms in any given ecosystem. Although Apple is best known for its computers and phones, its app store marketplace and distribution platform are among its most valuable and fastest-growing assets. The same is true of the Google Play mobile app store, Valve's Steam video-game distribution platform and gaming community, and Amazon's massive online retail platform, to name just a few. These organizations have had an outsize influence in shaping the ecosystems in which they operate.

The marketplace platform providers make money regardless of the size of any given transaction (for apps, games, music, and other content, platform providers typically take a 30 percent cut on every transaction). The cost of tools for creating digital content have dramatically dropped, and distribution has theoretically been democratized. But as Neil points out, the app ecosystem is more often famine than feast for the vast majority of content creators, with the bulk of the revenues going to the platform owners and the top content providers. Equally significant, these platform providers have become the gatekeepers for which apps, games, products, and designs are highlighted, usually through a blend of algorithmic (the most popular selections rise to the top) and curatorial (editorial selections) processes.

A critical question, therefore, is, who will control the primary platform for discovering, sharing, buying, and selling of fab designs, products, and services? The organizations that emerge in this role will shape the culture through the values built into their algorithmic and curatorial processes. The filtering in these processes is not neutral—every decision has built-in assumptions and there is considerable variation in the ways that community impact is integrated in the algorithmic and curatorial processes. An obvious example is the assumptions built into the Google and Amazon algorithms, which shape so much of the information we see and the purchases we make. Even if the ecosystem emerges more like cooking ingredients, as Neil suggests, there will still be gatekeepers—the equivalent of large supermarket chains that determine which products and services are highlighted.

Although we are starting to see a few companies providing marketplaces and online services for 3D printing or laser cutting, no dominant player or players have emerged. In the first two digital revolutions, the leading platform-based marketplace tended to be dominated by a few for-profit companies. Given the largely bottom-up, emergent nature of the fab and maker movements, the dominant marketplaces might emerge from the nonprofit side. Nonprofits like Wikipedia (making knowledge accessible) and Kahn Academy (making learning accessible) have experienced exponential growth and deep influence in their sectors, although they are more about knowledge and content than the processes of discovery, sharing, and selling of products and services. The organizations that emerge with the leading fab marketplace platforms will have great influence on the culture of the fab ecosystem.

Diverse Financing

Another key foundational component of an enabling ecosystem is the mechanisms for a diverse mix of private, public, and philanthropic funding for research and development, as well as stage-based financing for a mix of for-profit, nonprofit, and mission-based organizations across the fab community. For fab labs to maintain their exponential rate of growth, there will need to be significant funding for continued fundamental research, support for the growing network of community-based fab labs and the launching of new labs and networks, as well as impact-friendly commercial financing (angel, venture-capital, and private-equity) for companies developing fab hardware, software, materials, and services.

The fab ecosystem was born out of a research and development grant from NSF. On the roadmap that Neil envisions for digital fabrication there are still major technology challenges that must be overcome. For community fabrication to transition to truly personal fabrication, fundamental research is still needed. Traditionally, science and technology research has been funded by government institutions like NSF, the National Institute of Health, DARPA, and other government agencies (and their counterparts in other countries). And yet, we are living in a time when the efficacy of many core institutions of government, not to mention the value of fundamental research in science and social science, are being questioned.

Although innovation can and must happen in the private sector, underestimating the impact of public funding is not only dangerous, but is not supported by history. In her book *The Entrepreneurial State*, economist Mariana Mazzucato tackles the question of public funding head-on. She covers DARPA's role in the development of the Internet (through the creation of ARPANET). She also describes how government research funding in the United States and Europe laid the scientific foundations for everything from the touch-screen technologies powering smart phones to many pharmaceutical and medical breakthroughs. She highlights research funding by the National Institutes of Health and numerous other high-profile examples of fundamental research leading, years later, to products and services that have helped transform society.

Kumar Garg, who served for eight years at the White House Office of Science and Technology Policy, with a focus on education and innovation, suggested that "the key question for leaders in government is identifying the biggest pain points where the mechanisms of government or public-interest funding are best equipped to address." He also points out that, with limited financial and human capital, difficult choices often need to be made between a systems approach, where resources are used to address multiple pain points, versus a narrower approach, which starts by addressing the deepest pain point and prioritizes financial and human capital around that particular pain point.

For people interested in starting a fab lab, a common first question is, how much does it cost to launch a lab? Sherry Lassiter from the Fab Foundation estimates that the average budget for launching a community fab lab and running it for two years is approximately $250,000. The estimate includes hardware, software, consumable materials, space, and a mix of paid and volunteer staff to set up and run the lab. Funding for existing labs has come from a variety of sources: governments, communities, local businesses, universities, philanthropic organizations, investment, crowdfunding, and, most often, a combination of these. Ongoing funding has come from all these sources plus classes, fees for access, membership fees, and project grants.

The early cohort of fab lab pioneers are all passionate social entrepreneurs who have been able to cobble together both the start-up and the ongoing operational funding. For the network to continue to grow, however, the overall costs will need to come down, the variety of strategic funders will need to increase, and sustainable business models will need to be honed and adapted to meet local needs. A reduction in the overall infrastructure costs will require iteration and continuous improvement in hardware and software. A robust, enabling market ecosystem can make hardware and software better, faster, and cheaper and can support the innovation needed to ensure the availability of environmentally friendly and cost-effective raw materials.

One benefit of all the press around 3D printing is that it has inspired start-ups and attracted investment capital, which has resulted in competition, innovation, and cost reduction. We need to see this same market and investment support for subtractive hardware, consumable materials, and other areas where innovation and cost reductions are essential. This level of financial engagement will happen only if the products and services coming through the fab ecosystem become must-haves instead of just nice-to-haves.

Today there are individuals for whom having access to a fab lab is a must-have. Some of these people consider lab time a necessity because of what they can make in a fab lab. These include industrial designers and modern craft workers who require digital fabrication for their jobs. The same is true for some entrepreneurs, small businesses, and large commercial enterprises that need cost-effective, rapid prototyping. They must have the elements of a fab lab for the work they do.

For most of the fab community, however, the must-haves are more about the process of digital fabrication and being part of a community of makers and innovators. At this point in the development of the fab ecosystem, most people we spoke with said that the richness of the community (rather than the products being made) and the collaborative process of making is what makes fab labs so special.

Throughout the larger fab ecosystem, there is a tension between the process of making things with the digital fabrication tools and the resulting quality and usefulness of what is made. Maurice Conti, who directs applied research and innovation for Autodesk, a leading producer of digital design software, addresses both the enthusiasm for participating in fabrication and the limits on what can now be produced: "In terms of people able to do one-of-a-kind personal fabrication—that is super interesting, but is it practical? I have seen the way fab labs change culture—it is an infectious thing. Even here, with a world-class facility and expert staff, people are more excited about the process, with the output being rarely on par with commercial-grade goods."

It is one thing to learn how to laser-cut your name on a keychain or print a cute little 3D figure, but it is another to be able to make (almost) anything. Despite the great educational value in the process of learning digital fabrication skills, it will still be a long time before the average person can make most of what he or she consumes. Even Neil, who has access to the most advanced digital fabrication tools and technologies in his home and at work, makes only a tiny fraction of what he consumes. This will change over time as the technology continues to improve, but, in the interim, there is still great value in the process of making things in a fab lab and being part of the growing community.

The process of digital fabrication, when well scaffolded, cultivates critical thinking, design thinking, problem solving, creativity, collaboration, and resilience, as well a variety of digital literacies and growth in understanding of domains ranging from engineering to material science. For a growing number of educators—at least those who believe in learning by doing and collaborative project-based learning—this cultivation of knowledge, skills, and mind-sets are certainly must-haves. Similarly, the ways in which digital fabrication builds a motivation for and literacy in essential STEM skills is becoming a must-have for many educators, policy makers, and industries looking for future employees who can tackle a wide range of new challenges using rapidly developing technologies.

Haakon Karlsen, the "chief herder" of the Lyngen Alps fab lab in northern Norway, highlights the benefits of fab labs in the story of how three teenage girls who spent much of their free time in the fab lab all left town for university and went on to become doctors. People in the local towns would point to the Lyngen fab lab and comment how "that strange house on the hill in Lyngen makes doctors." Importantly, they came home after their medical training. And it isn't only doctors. Local youth who spent time in the lab have gone on to a variety of professional careers for which they credit their experience in the lab, from entrepreneurs to professors. For example, Hans-Kristian Bruvold, the boy who Neil described in Chapter 1, has recently become an associate professor at a local Norwegian university. Haakon was one of the first to recognize that fab labs cultivate a wide variety of knowledge, skills, and mind-sets that help people not only make (almost) anything but also become (almost) anything.

While these process-oriented benefits are great and need to be expanded to engage the broader population, must-have fab products and services will also need to emerge. There will be no single killer app or service that will define the fab ecosystem, but instead an evolving blend. In the early stages of community fab labs, the killer apps are likely to be about services that enable individuals to have deeply empowering experiences making things, rather than any specific thing being made.

As the fab lab network grows and we start to see early adopters for personal or small business fabrication, new killer apps and services will emerge—specific designs or projects that become must-haves to make money, save money, or provide great entertainment. These apps are analogous to the killer apps in the early days of the personal computer, such as word processing, spreadsheets, email, and computer games. Over time, the killer apps shifted from specific products to the enabling platforms and tools that democratized the creation of content on the Internet. These included tools for making web pages, blogs, and applications, as well as tools for uploading and sharing pictures, videos, news, and other information, which enabled the mass publishing and sharing of digital content.

Neil answers the must-have question, in part, by pointing to the personal part of fabrication. He argues that the killer app for digital fabrication is not what you can buy in stores but the ability to make what you can't buy in stores—products for a market as small as one person. To realize this vision, however, we will need to make great progress on enabling ecosystems and these, in turn, needed to be effectively governed.

Distributed Governance

Decision making and coordination are difficult in any organization. The complexity increases proportionately with distributed arrangements in which the parties are independent entities yet are still interdependent. Although often not an early priority with new technologies, effective governance across distributed communities is a critical element of an enabling ecosystem. Neil cites the governance of the Internet as an example of a distributed structure that is effective. Although Internet governance operates remarkably well given its scale and distributed nature, it has no authority or influence over critical issues such as access, literacy, civility, and other matters on a global basis. The challenge for digital fabrication is to advance decision making and coordination capabilities so that threshold challenges are addressed constructively, while still maintaining its distributed and independent nature.

At present, there is no shared platform for governance of digital fabrication in the ecosystem. The leading enabling practice—the fab charter—is incomplete. There is nothing wrong with the current language in the charter. However, it is missing key elements common to many charters, including a statement of the overarching shared vision, mechanisms for decision making, and mechanisms for resolving disputes. Moreover, the ecosystem lacks a full chartering process in which the charter truly represents an agreement among diverse stakeholders on the rules of the game. Here is the full text of the fab charter:

What is a fab lab?

Fab labs are a global network of local labs, enabling invention by providing access to tools for digital fabrication.

What's in a fab lab?

Fab labs share an evolving inventory of core capabilities to make (almost) anything, allowing people and projects to be shared.

What does the fab lab network provide?

Operational, educational, technical, financial, and logistical assistance beyond what's available within one lab.

Who can use a fab lab?

Fab labs are available as a community resource, offering open access for individuals as well as scheduled access for programs.

What are your responsibilities?

- Safety: not hurting people or machines
- Operations: assisting with cleaning, maintaining, and improving the lab
- Knowledge: contributing to documentation and instruction

Who owns fab lab inventions?

Designs and processes developed in fab labs can be protected and sold however an inventor chooses, but should remain available for individuals to use and learn from.

How can businesses use a fab lab?

Commercial activities can be prototyped and incubated in a fab lab, but they must not conflict with other uses, they should grow beyond rather than within the lab, and they are expected to benefit the inventors, labs, and networks that contribute to their success.

In our survey of fab leaders, the majority (69.8 percent) reported that awareness and utilization of the fab charter is important, but approximately 11.3 percent felt strongly that it was not important. Furthermore, 56.0 percent of respondents reported that it was difficult to utilize. Most tellingly, a significant percentage, over 20 percent, responded to both questions with either "Don't know" or "Not applicable" (indicated by brackets).

When we shared these findings with Neil, he observed that the charter was essential early on, but is less important now since the shared understandings are embedded in the way the Fab Academy operates and other parts of the fab landscape. In our interviews and survey, we have indeed observed a set of shared fab values across the community. To the best of our ability, we have summarized these as follows:

- Finding meaning, purpose, and joy in making in a fab lab
- Enabling collaboration and community building within and across fab labs
- Cultivating individual and collective agency in the digital fabrication process
- Working toward personal and societal sustainability through digital fabrication
- Digital fabrication should be benefit everyone, not just the fortunate few

These shared values across the current ecosystem do seem to be serving the same function as a charter. Indeed, these shared values go beyond what is stated in the current charter, and some elements of the charter conflict with current practice. For example, under "Who can use a fab lab?" the standard is that fab labs offer "open access for individuals as well as scheduled access for programs." This standard is not always possible for fab labs in K–12 schools, colleges and universities, museums, and other institutions for which there may be constraints on open access.

Still, just stating the shared values raises the question of whether some codification—an updating of the charter and process to ensure that

FAB CHARTER

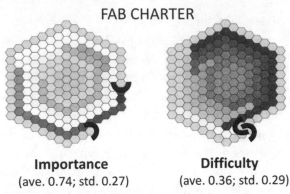

Importance	**Difficulty**
(ave. 0.74; std. 0.27)	(ave. 0.36; std. 0.29)

My awareness and utilization of the fab charter (.38 gap between the two means).

Importance and difficulty of access and utilization of the fab charter. *WayMark Analytics*

it is shared—is needed as the ecosystem grows. Particularly in an age of accelerating change, charters need to be living documents. A mind-set for this can be found in the US Constitution and other chartering documents that provide an enduring framework, combined with the capacity to adapt. Constructing such mechanisms is not easy but represents a crucial codification of how things are meant to operate in a given domain. A challenge going forward is to ensure that the enabling platforms, practices, and people who help with governance—all part of the ecosystem—can handle the exponential rates at which things will change in the third digital revolution.

MITIGATING RISK

Technology enthusiasts will acknowledge risk, but their enthusiasm centers on the beneficial potential of the technology. This drives innovation, which is good. However, a clear-eyed, early look at the third digital revolution also reveals considerable risks that need to be documented, understood, and addressed (even if they can never be fully solved). The issues of fab access and literacy, for example, surface the risk of a potentially destructive fab divide. Other risks include environmental degradation associated with consumable materials, misuse of the technology (e.g., bad people making bad things), and unintended consequences that come with people's ability to manipulate bits, atoms, and genes at an increasingly granular level.

The two overarching institutional objectives in society are creating value and mitigating harm. A central challenge in mitigating harm focuses on who (or what combination of stakeholders) will hold responsibility for each risk. For any large commercial or public project, there is always a crucial negotiation around who will hold which risks. The process begins with the identification of all known risks and the broad categories from which unknown risks might emerge. Then, if the negotiation goes well, each risk is allocated to the party or parties best able to manage a given risk. The person or group receives payment or other compensation for holding the risk. If these negotiations can't reach resolution, the project is not likely to go forward. A current example of this is the negotiations around who holds the risk for self-driving cars: the car manufacturers, the owners of the cars, the insurance companies, the development teams who wrote the algorithms, the government, or others.

For the third digital revolution, however, there is no prescribed mechanism for negotiating with multiple stakeholders on risk. Thus, things may advance, but the responsibility for managing risk is incompletely specified. The failure to systematically assign responsibility poses what might be considered a meta-risk for the entire third digital revolution. In the absence of risk-mitigation agreements, key risks will be left unmanaged, and when catastrophic events do occur, there will be inadequate responses and resources, with the further risk of contentious finger-pointing making things worse.

Begin with environmental risks. The current hype around 3D printing, has some pundits predicting "a 3D printer on every desk and in every home," paraphrasing the original Microsoft mission statement. Considering current technology, this goal could be an environmental disaster. The 3D printing process is basically the smooshing together of various materials. Even though one common material, PLA, is plant-based, others are environmentally unfriendly—derived from petrochemicals—and can't be easily recycled.

As Tomas Diez from the Barcelona fab lab says, "fab labs are using more and more plywood with glue, plastics that use ABS [acrylonitrile butadiene styrene], and electronics that depend on acrylics. This is definitely not sustainable." When hundreds of millions of people are iterating and tinkering with environmentally unsustainable materials, the potential environmental risks of the third digital revolution include a severe depletion of natural resources, a dramatic rise in carbon emissions (from the development and shipping of these materials), and the excessive waste currently generated in the digital fabrication process.

And yet, if we can develop environmentally friendly and reusable consumable materials for digital fabrication, there could be great environment dividends. In the Fab City white paper, Diez points out cities' current role in environmental degradation: "Extreme industrialization and globalization have turned cities into the most voracious consumers of materials, and they are overwhelmingly the source of carbon emissions through both direct and embodied energy consumption; we need to reimagine the cities and how they operate." As we will see in later chapters, the development of better resources on the use and reuse of local materials and the advent of digital materials that can be assembled and disassembled will be essential to avoid the environmental risks of universal personal fabrication.

Another risk in developing countries derives from weak institutions. Multinational corporations will be able to establish digital fabrication capabilities faster than many local institutional leaders. This is already beginning, with the United Parcel Service setting up rapid prototyping centers in its warehouse facilities to fabricate and deliver certain products on demand. Here the risk is that the gatekeepers to the technology will not be distributed community centers, but will be primarily large corporate enterprises that are more analogous to what has happened with the first two digital revolutions. In this scenario, the third digital revolution will come more rapidly to the developing world, but the type of change will be mediated by what is and is not profitable for business (which may or may not be aligned with fab values).

There is also the more sinister risk that ill-intentioned people will make bad things. It is already starting to happen. Stories of 3D-printed guns are quickly become viral accross social media, making multiple laps around the Internet. The fabrication of weapons is a real risk. Neil responds to this concern by pointing out that fab labs are not the most efficient way to make guns—that there are much more efficient ways to procure a gun than building one in a fab lab. His argument, however, becomes less persuasive as the tools for digital fabrication get better, faster, and cheaper. If the goal is to enable anyone to make (almost) anything, then certainly this goal will accelerate the making of dangerous and destructive things.

There is no failsafe way to solve this problem, but there are multiple ways to help mitigate the risks. One element of mitigation is through a combination of mentorship and oversight at a local lab level. Yogesh Kulkarni, from the fab lab outside Pune, India, has actively addressed this challenge. He believes fab labs can be an important tool to help identify potentially at-risk youth and guide them toward more productive paths. When asked about the risk that these same youths would use the lab to make weapons, he said that close oversight is key. "Although the risk is small, we will, for example, see a student use the grinder to sharpen a knife. When this happens, we stop them and explain that is not what the lab is used for." This approach is similar to advances in community policing, where small early incidents are caught in a systematic way—to set a tone and culture that makes the larger, more catastrophic events less likely.

A related risk at a systems level is cyberterrorism in a world of digital fabrication and, even more concerning, a world of digital materials. This points to the need for distributed infrastructure without single system-wide failure points. Moreover, we know that system solutions will have to be social, as well as technical.

Although Alan and Joel have spent their careers in very different sectors, both have insights on innovative approaches to managing technological risk. These insights come from their respective domains—workplace and labor practices (Joel) and digital entertainment and online communities (Alan). This speaks to the importance of cross-sector collaboration and knowledge sharing as virtually every sector and domain is having to address a steady stream of risks introduced by accelerating technologies.

We'll start with Joel. He was on one of two teams that the National Aeronautics and Space Administration (NASA) engaged to conduct safety and risk assessment of a prospective Moon-Mars mission. One team used what is called *probabilistic risk assessment (PRA)*, which counts all known possible failures, attaches a probability to each one, and then calculates the overall risk of the mission. The other team, led by MIT's Nancy Leveson, used a *system safety model*, which looks at the operation of the overall system and identifies points of vulnerability according to a systems dynamics model of interdependencies. One approach decomposes the risks into their component parts, and the other looks at how things add up and interconnect with each other. Unfortunately, risk analysis is too often focused just on the PRA-style decomposition of risks, with less attention to the systems approach. In this case, Joel recalls seeing the introduction to the PRA report which stated that it was a complete assessment of all known risks associated with the mission—*except those associated with software and organizational factors* (italics added). Needless to say, risks in the third digital revolution can't only be addressed by decomposing the risks for the individual component parts.

Beyond formal risk-mitigation processes, some legal aspects of risk are also a threshold challenge. Because some physical objects are critical to safety, fab labs must ensure that such objects are what they say they are. A counterfeit or even a well-meaning substitute component could subvert a product's design for safety. The challenge involves issues of authentication and intellectual property. Gonzalo Rey, who was the chief technology officer for Moog, which makes precision control parts for aerospace, medical, energy, and other sectors, discusses authenticity: "Our company does worry about authenticity. People like to have a brand-name watch, for example. What is urgent for us is that metal printing is becoming real. If you are doing things like an airplane or a surgical part, you want to know that it is the authentic part. This is in the same technical bucket as the designer watch—involving the digital rights that go with the design. In safety-critical situations, it is essential."

Another risk that is already emerging around new technologies involves labor practices. Much of the work in the current digital fabrication ecosystem is volunteer or part-time contingent work referred to as gig economy jobs. The emerging mechanisms to share designs, products, and services commercially also involve fragmented work. This is part of what labor economist David Weil terms the "fissured economy," where policy objectives of fair pay, workplace safety, prevention of discrimination, opportunities for representation, and related protections are getting harder to reinforce. Here the risk today is that the work in fab labs will be part of the larger current societal challenge with fragmented work. This risk will vary across the settings in which digital fabrication takes place—schools, colleges, universities, museums, community labs, industrial settings, and others—with some settings involving more fragmented work than others.

In time, there is a greater risk around which labor and workplace practices will emerge to characterize the ecosystem. It is possible the collaborative and other constructive practices will be institutionalized, which would be excellent. The risk is that arrangements that increase power differences and reduce collaboration will dominate instead. For example, the platforms for the sharing of designs and the exchange of products and services could end up with a small number of owners for whom the platforms are highly profitable, with limited benefits for the many millions of contributors. Such a situation would be a barrier to a more self-sufficient and sustainable society.

The behavioral strategies associated with making dangerous or destructive things are part of a larger risk-mitigation approach centered on building in norms that cut across all digital communities—addressing critical issues around digital civility and mitigating toxic behaviors. An interesting example of how to do this comes from an unlikely source, the video-game industry. Racism, homophobia, sexism, and general toxicity have become pervasive in some of the larger online gaming communities. Game developers have had to tackle this problem head-on for both business and ethical reasons. A company proactively addressing this challenge is Riot Games, developer of the hugely popular *League of Legends* game. With close to seventy million players, the community is massive and global. Riot Games found that toxic behavior was frustrating many existing community members and scaring off new players. Since early design priorities did not prioritize the potential for these antisocial behaviors, the company had to rethink some of its core community and design assumptions.

Riot Games engaged social scientists, neuroscientists, and game designers to study the problem. They studied in detail who in the community was driving the toxic behavior as well as how this behavior spread and how it influenced the community. They also studied the same factors in positive behavior. The researchers then openly experimented with ways to integrate mechanisms into the flow of the game to reduce the toxic and increase the positive behaviors. The effort included various initiatives. The company tried adding peer reviews to the game, to help determine if particular behaviors were toxic. Another initiative was the offering of incentives (along with messages and tips) to encourage positive behaviors among the gamers. The company also ensured that the penalties for different types of objectionable behavior were aligned with the nature and severity of the transgressions. They also used machine-learning algorithms to track progress in real time across multiple languages and cultures. At the core of this approach was the clear, specific, and nearly instant feedback given to the player when toxic behavior happened (doing nothing both condones and reinforces toxic behavior). Equally important was a mix of transparency and humility (not overstating what was intended) in terms of the process and carefully cultivating the better "angels" of the community.

The results have been impressive. When *Nature* magazine interviewed Jeffrey Lin, the lead designer of social systems at Riot Games, he highlighted how an in-game warning that harassment leads to poor in-game performance "reduced negative attitudes by 8.3 percent, verbal abuse by 6.2 percent and offensive language by 11 percent compared with controls." A positive message about players' cooperation reduced offensive language by 6.2 percent. And after the company released its machine-learning algorithm, verbal toxicity relative to other ranked games (the most competitive level among video games) was reduced by 40 percent.

Riot Games has said it is going to publish its results so that others can leverage these insights. In this example, the software developers were willing to "hold" the risk since it was affecting their business model and their customer experience.

Building motivation for positive behavior into the design and continual optimization of the software is just one example of many possible built-in technological solutions to managing risk. A key aspect of the science underlying digital technologies involves the concepts of error correction, which Neil explores more fully in the next chapter. Error correction is more difficult with people and societies, but the Riot Games example indicates some of the ways in which it is possible. Thinking along these lines, machines might someday sense certain misuses of materials or might apply artificial-intelligence (AI) learning when new forms of misuse emerge. In the development of such approaches and responses to possible misuse incidents, people will still need to make decisions. Thus, an essential part of any risk-mitigation strategy involves having a broad array of forums and mechanisms dedicated to tracking and addressing issues when they arise, combined with anticipating what might be on the horizon and then combining the social processes with the technology.

While the Riot Games example illustrates the use of technology to nudge behavior toward more civil forms of discourse, it assumes a threshold acceptance of the technology itself. Betty Barrett, a co-founder of the Champaign-Urbana Community Fab Lab and a co-leader for the research team examining stakeholder alignment among US fab labs, points to a deeper risk: "Exponential change is not always onwards and upwards. Rates of change can go backwards. There are forces in the world that see it in their interest to head backwards. This includes fundamentalist beliefs and anti-intellectualism." She adds, "The intense disbelief and distrust in science limits the rates of change. This is a huge battle and digital fabrication will be subject to the terror that is felt by many around technological change." There is some irony, of course, that these constraining views are magnified and accelerating in their impact because of technology. Barrett notes two risks here. First, digital fabrication is not ready to face what could be violent ideological headwinds that find its very existence offensive. Second, some parts of the world will not face such strong resistance and will race ahead—deepening the fab divide.

These challenges rooted in fear could deepen as we reach the later stages of the third digital revolution and as new risks emerge that are more opaque and even existential. Neil often talks about how machines will make machines and, as we will learn in future chapters, assemblers assembling assemblers—but it is not clear where humans fit into these evocative descriptions. How does the process start and stop? Does it stop? Could it run amok? When people hear about machines making machines and assemblers assembling assemblers, it is hard for them not to leap to visions of rogue AI wreaking havoc on humanity. When we consider the many existential threats to humanity, the inevitable interweaving of AI with digital fabrication is certainly on the list. Here the threshold challenge begins with cultivating an informed and engaged population around the future of the third digital revolution. The technology needs to be presented in a way that is inviting, highlights clear and understandable capabilities within reach, and addresses genuine fears. As we will explore more deeply in Chapter 6, we must simultaneously appeal to the head (logic-centric approaches), the heart (emotion-centric approaches), and the hands (practical hands-on approaches). This is to ensure that social systems can co-evolve effectively with the accelerating technologies.

The potential of the third digital revolution also raises questions about who in society will be the arbiters of the values that are embedded in the technology. Will we trust the technologists, or will we also engage ethicists, social scientists, storytellers, religious leaders, and mediators? We can (almost) envision processes for addressing the moral issues associated with digital technology, but it is more difficult to have confidence that the results of these processes will be seen as legitimate and will prove durable.

Although fab labs must mitigate their own risks, they can also play a role in reducing risk in society. As Neil introduces in Chapter 1, Dina El-Zanfaly co-founded a fab lab in Cairo around the same time as the January 2011 Egyptian revolution. In fact, it is located not far from Tahrir Square. Amid the tumult of the protest, the fab lab turned out to be a neutral, safe place. El-Zanfaly recalls that it was "a bubble of positive energy and enthusiasm. It was somehow related to the revolution, but it was also separate." Even as digital fabrication brings new risks, fab labs also foster community and innovation in ways that can reduce risk at times of protest and change.

Another area where digital fabrication capability can reduce risk was articulated by US Representative Bill Foster. Focusing on national security, Foster points out that there are military and natural disaster threats that could disable substantial parts of the economy. In these scenarios, he points to distributed fabrication as an essential capability for rebuilding a well-functioning society after catastrophes. Foster further describes a military scenario that involves the increased use of robots in warfare. Under this scenario, an invading power could position militarized robots on every street corner. The ability to fight back might well hinge on local and national digital fabrication capability to generate rapid prototype responses to unexpected technology threats.

Foster has proposed a bill in Congress, indicating the value of a national fab lab network. He recognizes the many benefits of fab labs as sources of learning, community, and innovation, in addition to national security. But for many in Congress, attention increases when the emphasis is on a network of labs with the potential to mitigate catastrophic risks in society.

GRAND CHALLENGES

Society faces many grand challenges, each with global risks attached. The United Nations has a list of seventeen "sustainable development goals," including many that can be advanced by the third digital revolution: no poverty; sustainable cities and communities; responsible consumption and production; decent work and economic growth; reduced inequality; industry, innovation, and infrastructure; quality education; and peace, justice, and strong institutions. Neil even ran a fab lab at the launch of the goals. There is a reason that foundations, professional societies, and global organizations are all articulating grand challenges. It is because most of society's institutions are not set up to effectively tackle these complex, and continually evolving, cross-cutting issues. As economic historian and Nobel Prize winner Douglas North defined them in 1990, institutions "are the humanly derived constraints that shape human interaction." More simply, they are the rules of the game, and digital fabrication has the potential to be a game changer for societal institutions. Realizing this potential, however, involves first navigating the inherited institutional context.

Many existing institutions, such as those associated with education, business, communities, and the environment, are struggling to keep pace with accelerating technologies. For the most part, the existing institutions represent sources of conservatism that pose potential barriers for something new like digital fabrication. The third digital revolution will simultaneously engage the existing institutional actors and reveal cracks in the system—cracks that will be exploited by the emerging, more agile people, platforms, and processes.

In our survey, we asked the fab pioneers to look ahead to the year 2025 and anticipate the degree to which digital fabrication will transform key elements of society—education and training, business and entrepreneurship, government and civic organizations, environmental and economic sustainability, and community and culture. Of this group, many had roles that connected to these institutional arrangements. With respect to design and fabrication, 64.7 percent reported having educational roles, 53.9 percent reported community organizing roles, and 35.9 percent reported public leadership roles. The accompanying figure shows the pioneers' average responses about how great an impact digital fabrication would have on various aspects of society (on a numerical rating scale, where 0 means "not at all" and 10 means "complete transformation").

Given that 2025 is less than a decade away, those closest to digital fabrication technologies anticipate considerable transformation, with the most change anticipated in educational institutions and practices, and the least in government and civic organizations. Understand that these are the pioneers' predictions, not what they desire. The point of prediction is not to describe what you want, but rather to anticipate where things are headed and figure out where to intervene or how to act now to accomplish shared objectives. Thus, a lesson from this predictive exercise is that those closest to the new technology expect that government and civic organizations will be lagging in the technology. The prediction should be taken as

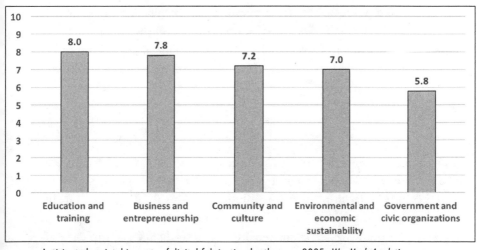

Anticipated societal impacts of digital fabrication by the year 2025. *WayMark Analytics*

a challenge for these institutional leaders to lean forward and, in effect, prove them wrong.

As Neil noted in Chapter 1, the UN leaders did not see an immediate connection between grand challenges and digital fabrication. But when they did, their whole attitude toward digital fabrication shifted from considering it a nice-to-have to begin viewing it as a potential must-have. For the third digital revolution to truly become a revolution, it will need to do more than demonstrate exponential gains in technical performance. The technology must clearly address enduring societal needs with models that can be replicated and locally adapted for broad-based impact. Just because there is a conceptual fit with the UN sustainable development goals doesn't mean that it is the reality on the ground.

As Tom Kalil, former deputy director of the White House Office of Science and Technology Policy and a strong advocate of the maker and fab movements, points out, "the ability of the maker or fab communities to help solve particular societal problems is a key yardstick to measure progress. It is not just about self-expression." He goes on to cite examples such as makers and people with disabilities working together to design assistive technologies, and makers and educators working together to reduce dropout rates in schools.

Beyond progress on tangible impacts at the level of societal grand challenges, there are also risks involving intangible impacts. Derrick Mancini, who served as project director for the design and construction of the Center for Nanoscale Materials at Argonne National Labs, observes: "People are drawn to digital fabrication as a reaction to the accelerating changes in technology. There was a craft movement in response to the industrial revolution, and the draw to the fab and maker movements is comparable—when people reach out to maker technology, it reveals a deep desire for individual control." In this way, the rise of digital fabrication has great potential for people to achieve greater personal control at a time of accelerating change.

▪   ▪   ▪

To summarize, woven through the mix of exhilaration and frustration that exist in the fab ecosystem are the four threshold challenges central to the future of the third digital revolution: universal fab access, fab literacy, an enabling fab ecosystem, and mechanisms to mitigate potential risks. There are also persistent tensions such as balancing the ease-of-use of the technology and avoiding black-box solutions; reconciling the need for interoperable standards and protocols with the distributed and independent nature of the community; and balancing nonhierarchical decision making across diverse stakeholders with the various needs for collective action. Our combined ability to meet these challenges and manage these tensions will likely determine whether we can tip the balance toward an optimistic vision of the future where all people can make (almost) anything—or a future where some people can (almost) make anything.

CHAPTER 3

The Science

To see where digital fabrication is heading, we first need to understand where it's coming from. Depending on how you count, the third digital revolution could be recognized as starting around 2000 with the maker movement and the first fab labs. Or it could be considered to have started around 1950, when an early computer was connected to a manufacturing machine for the first time at MIT. Or its start could be dated to around four billion years ago, when life evolved the machinery for molecular manufacturing.

Whichever of these you choose, there's a close historical parallel with the corresponding stages in the digitization of communication and computation. All three digital revolutions are based on the same core ideas. And the common historical lesson is that the implications of each could be seen and used long before the revolution was apparent and the technology reached its final form. The past is indeed the prologue for the third digital revolution, as we'll see in this chapter, beginning with the projections that originally led to Moore's Law and are now leading to Lass' Law.

FROM MOORE'S LAW TO LASS' LAW

In 1965, Gordon Moore published the most famous graph in history, the one that has come to define the digital revolutions in communication and computation. *Electronics* magazine had asked him to forecast the use of electronic components for the next decade. At the time, Moore was the head of R&D at Fairchild Semiconductor, and he went on to co-found Intel in 1968.

He saw the future in five data points. When the number of components in an integrated circuit was plotted against time, the results showed that they were doubling every year or so. The classic reason for doubling is reproduction. After a cell divides, there are two cells. Then after those divide, there are four, then eight, then sixteen, and so forth. This series can be written with an exponent counting the number of factors of 2 to multiply: $2 = 2^1$, $4 = 2 \times 2 = 2^2$, $8 = 2 \times 2 \times 2 = 2^3$, $16 = 2 \times 2 \times 2 \times 2 = 2^4$,. . . . The logarithmic function does the opposite of the exponential, counting the number of multiples: $\log_2 2 = 1$, $\log_2 4 = 2$, $\log_2 8 = 3$, $\log_2 16 = 4$,. . . . That's why a linear increase in a logarithm (1, 2, 3, 4) corresponds with an exponential increase in the quantity being measured (1, 2, 4, 8). And it's why Moore plotted the logarithm of the number of transistors against time, so that the doubling trend would appear as a straight line.

Many of our physical senses have a logarithmic response. Because we perceive a doubling of the intensity of a light or of the volume of a sound as equal increments, we can go from a dim room into bright sunlight without being blinded, or from a quiet hallway into a loud concert. But our perception of the world is more typically linear. Bank statements count the number of dollars, not the number of zeros, in an account; clocks measure the passage of time, not the doubling of time. With just five data points, the logarithmic plot doesn't look very different from the linear plot. But the relationship is called Moore's Law rather than Moore's Graph because he made the leap to extrapolate that the straight-line trend would continue for a decade. If you plot the number of transistors on a linear graph, it appears that nothing is happening until there's

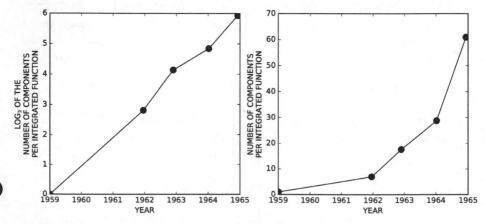

Gordon Moore's initial data, on logarithmic and linear scales. *Neil Gershenfeld*

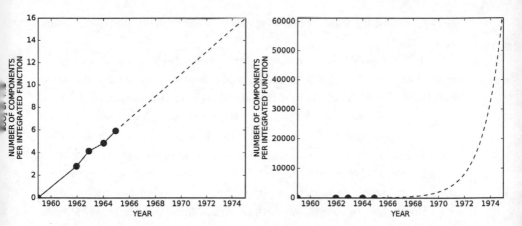

Moore's Law initial projection, on logarithmic and linear scales. *Neil Gershenfeld*

an abrupt revolution, but the logarithmic plot lets your eye see that the number is steadily doubling.

Moore almost got this right; the doubling actually continued for five decades. The heart of Moore's Law can be seen in the number of transistors in the microprocessors that Intel produces. Over the following four decades, the number has continued to double (extending the straight line on a logarithmic plot). If you plotted the numbers linearly, you would probably conclude—unless you were paying attention—that nothing much was happening for decades and that around the year 2000, a revolution in digital technologies occurred. That might have been how it felt with our linear

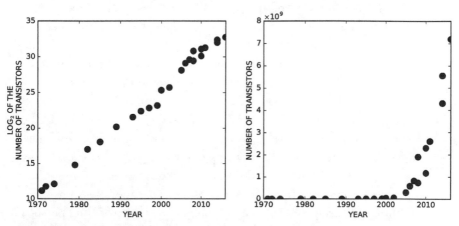

The number of transistors in Intel's microprocessors, on logarithmic and linear scales. *Neil Gershenfeld*

perception, but the logarithmic plot makes clear that the trend was going on for decades.

What is reproducing in an integrated circuit? The best answer is an indirect one: the bits are. The analog alternative doesn't behave the same way. In the book's introduction, we observed that if a page is fed into a copier, there are then two copies, and then if those are copied, there are four, then eight. But each successive copy degrades slightly, until subsequent copies all end up a garbled mess. Doubling can continue only if it's coupled with error correction. Cells do that by detecting and correcting errors in the genome when it's duplicated to be passed on during cell division. The biggest integrated circuits now have billions of transistors; each one degrades the signals, but there's an exponential reduction in these errors because the circuits keep restoring the logical states of the bits. This ongoing error correction can keep up with the exponential increase in transistors, so that the output of the billionth transistor is as reliable as the first.

Transistors don't literally make transistors, but metaphorically they do. They go into chips that go into computers that are used to design better computers; they are used to send messages that prompt more messages; they are used to write programs that are used to write programs. The result of this feedback loop has been an exponential improvement in not just the number of transistors that can fit on a chip, but also in their size, speed, and cost. These improvements have led in turn to exponential improvements in digital communication and computation systems, including the size of memories, the speed of networks, and the number of connected computers. Since Moore never actually called his observation Moore's Law, there has been some debate about exactly what Moore's Law measures. The best answer is the performance of information technologies. The importance of Moore's Law comes not from its definition, but rather from the observation that decades of data points have followed the initial trends.

When plotted against time, each of these measures of the performance of digital technologies forms a straight line on a logarithmic scale, but that doesn't mean the embodiments of the underlying technologies have been straight lines. Instead, they appeared in epochs. The first messages sent via the precursors to the Internet were between building-filling mainframes. When the Internet reached one thousand sites, they were room-filling minicomputers; at one million, they were desk-filling personal computers; and at one billion, they were pocket-filling smartphones. And the same kind of thing was happening inside the devices in these systems.

Integrated circuits were first made by manually cutting out the artwork for exposure masks that photographically define where to etch and deposit material on a silicon wafer that will become an integrated circuit (a lost art that I originally learned). This step was then automated. The masks were first placed in direct contact with the wafer and later projected with reduction optics. The exposure was first done with visible light and later with shorter-wavelength extreme ultraviolet light, and so forth. In each of the decades of Moore's Law, there were confident predictions of the imminent demise of further improvements because of looming difficulties in any one of these technologies; each of these technological transitions allowed the performance scaling to continue.

Exponential growth does eventually reach resource limits. The technical term for that is a *sigmoid function*, which just means an S-shaped curve. There is a period of what appears to be flat growth (the bottom of the S), then a period of accelerating growth (the middle of the S), and then a return to slow growth as constraints are encountered (the top of the S). In biological cell growth, the cells run out of space and nutrients. Moore's Law is running out of physics, as it hits fundamental limits. Around 2015, all the quantities that had been doubling began slowing down. The features in mass-produced chips are getting so small that it's necessary to keep track of the individual atoms in the devices, and in the lab there are single-electron transistors that do just that, turning on and off depending on the presence or absence of one electron. A transistor cannot shrink any smaller than that, unless we start programming atomic nuclei, something that's not done outside of high-energy particle accelerators.

Moore's Law is also running out of people. Once almost everyone has access to a computer and is connected to the Internet, there's no longer the same demand that drove these devices' development. Digital computing and communications are becoming commodities that are incrementally replaced rather than rapidly adopted. If people are like the nutrients feeding the spread of computers, the sigmoid curve of adoption starts slowing down well before it hits its ultimate limit, which is why, as Joel and Alan explain, reaching the last few billion humans is taking so much longer than did reaching the previous few billion.

The end of Moore's Law is the source of great angst for both technologists and economists, because it corresponded to a period of tremendous growth in productivity and prosperity. The assumption is that if the technology stops improving, so will the economy. It's less of a concern for most consumers, who are now bombarded by a sea of digital data competing for their attention and who are no longer so obviously running into performance limitations in their computers.

Saturating demand runs into what has been called Moore's Second Law: the cost of chip fab. He didn't originally plot this, but the cost of the factories that make state-of-the-art chips has been growing along with everything else, so that these facilities now cost billions of dollars and the development of a current-generation chip can cost a hundred million dollars. Once the money is spent, the transistors cost fractions of a cent, but the huge capital outlay has become prohibitive for all but the biggest manufacturers and markets.

The cost of a chip fab has gone up rather than down because it's based on analog processes. Materials are continuously deposited or etched, extreme demands are placed on the uniformity and tolerances of the processes, and any defects can ruin a chip. One of the most closely guarded secrets in the industry is the yield—the depressingly small fraction of good chips made in a state-of-the-art process.

The promise of the third digital revolution is to move manufacturing from Moore's Second Law to the first one. Around when we set up the first fab lab in 2003, I visited Gordon at Intel to consider this possibility. He arrived at our meeting from the kind of modest cubicle that all Intel executives work in. He is very humble about his role; he didn't name the law after himself, and he had a hard time referring to it that way—the name came from a Caltech collaborator, Carver Mead. Intel's keepers of Moore's Law maintain a clear organizational separation between the computer scientists developing the programs and logic for integrated circuits, and the physical scientists developing the devices and process technologies to build them. The computers are digital, but the manufacturing processes are analog.

As described earlier, I first appreciated that there might be something like a Moore's Law for digital fabrication when Sherry Lassiter, manager of the fab lab project for CBA and later the Fab Foundation, noticed after a few years that the number of fab labs was doubling roughly every year and a half. Hers was a decidedly analog initial observation: the pile of papers on her desk with fab lab correspondence appeared to be doubling in height. We jokingly noted the parallel to Moore's law, what we are calling here Lass' Law. Like Moore's Law, the projection has already extended much further than we anticipated. After a bit more than a decade, the number is around a thousand fab labs, the result of ten doubling cycles. Here the origin of the reproduction is clear; once a fab lab opens,

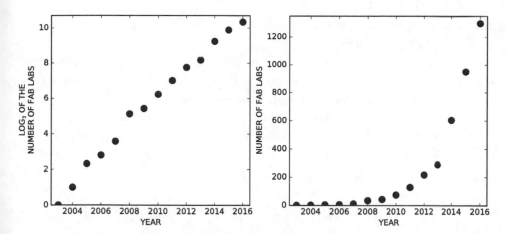

The number of fab labs, on logarithmic and linear scales. *Neil Gershenfeld*

it inspires people who see it to want another one. And what could be considered error correction at this stage is how the fab labs have an evolving set of the same core capabilities so that people and projects can be shared.

The first point in the plot of fab lab growth starts in 2003 with the first fab lab, rather than in the 1950s with the invention of computer-controlled manufacturing. In the 1950s, the size and cost of a fab lab would have been matched by each of the ten or so computers required to run all the machines. Not until (thanks to Moore's Law) the cost, size, and performance of digital computing and communications could be contained within each of the machines did fab labs become feasible.

As with Moore's Law after its first decade, it's now both ambitious and prudent to project that the doubling of fab labs will continue after their first decade. That means the equivalent of a million fab labs after the next decade or so and a billion following the decade after that. Like the spread of the Internet, Lass' Law doesn't mean a billion room-filling facilities. Rather, each of these decades marks a technical epoch in the integration and accessibility of the technology to make (almost) anything.

Like the many quantities that came to be subsumed by Moore's Law, Lass' Law is likely to extend to a number of ways to characterize the performance of digital fabrication. These measures include the number of parts placed, the rate at which they're placed, the size and cost of a part, and the complexity of what can be assembled with them. When the Internet reached a billion computers, the chips in those computers were reaching a billion transistors; when there are the equivalent of a billion fab labs, they'll likely be making things that contain a billion parts. Believing in that is as absurd as believing in Moore's Law. In Chapter 5, I'll look at how we're going to get there.

Lass' Law could be viewed as a continuation rather than a replacement for Moore's Law. Moore's Law has lived in flatland, the real estate on an integrated circuit; Lass' Law lifts it from 2D into the 3D world we live in. And Moore's Law has applied to the specifications of chips after they're produced; Lass' Law extends it back to how they, and everything else, are made. The conclusion of the third digital revolution is that the long-sought killer app for the future of computation is fabrication.

COMMUNICATION, COMPUTATION, FABRICATION

Moore's Law can be viewed as being like what an economist would call macroeconomics, or what a physicist would call thermodynamics, an aggregate system property. To understand the workings of the historical parallel between Moore's Law and Lass' Law, we need to look beyond these scaling graphs and examine the equivalent of microeconomics, or what a physicist would call statistical mechanics, the individual elements that make up the system. These are the computing and communicating machines in the first and second digital revolutions and the fabricating machines in the third digital revolution.

When computers were first developed by and for large institutions, they filled buildings. The architecture of modern interactive computers can be traced back to the first major computer that could respond in real time to inputs as they happened, rather than by processing them in batches. The Whirlwind I was developed by MIT's Servomechanisms Laboratory, originally to replace the analog computers that ran flight simulators. It later became the basis of the semi-automatic ground environment (SAGE) air defense system. Developed in 1947, the Whirlwind I became operational in 1951. It cost several million dollars, and filled a few floors of a building.

The first computerized control of a manufacturing machine was an offshoot of this project, in 1952. An inventor named John Parsons brought to MIT a proposal to connect a real-time computer to a milling machine to make the increasingly complex parts required for new jet aircraft. The Servomechanisms Laboratory embraced the idea, but not him, leaving Parsons off the Air Force project developing what we now call a numerically controlled (NC) mill. A computer could continuously move the rotating cutting spindle in all three directions simultaneously, following paths

The first computer-controlled milling machine at MIT. *Courtesy of MIT Libraries, Institute Archives and Special Collections, J. Francis Reintjes Papers, MC-0489, Box 8*

that would be impossible for a machinist to do with two hands. Many kinds of fabrication tools have since been used in the machine's descendants to replace the rotating spindle of the milling machine, but the basic principles have otherwise been unchanged since 1952.

The successor to the Whirlwind I at MIT was the TX-0 and then the TX-2, the first significant computers to use transistors rather than vacuum tubes. This difference is important in a number of ways. First, the transistors could be packed into a smaller space. Second, they did not generate the same amount of heat when operating. Third, they required less power to operate. Fourth, they were more reliable. Together, these benefits enabled a smaller, more efficient device that could make faster calculations.

The TX-0 and TX-2 were first operational in 1956, and the engineering team led by Ken Olsen spun them off a year later as the PDP (programmed data processor) family of computers sold by DEC. PDPs brought the cost of computing down from millions to hundreds and then to tens of thousands of dollars, and the size of a computer down from a building to a room and then to an equipment rack. PDPs weren't easy to use—when I learned how to use one, it was necessary to connect and master separate units for processing, memory, storage, input, output, communication, and power. But because they could be owned by a workgroup rather than a whole organization, someone like me could get access to one. Just about everything you now use a computer for today was first done on a PDP—from writing documents to playing video games to sending messages over what became the Internet. DEC went on to spawn a computing industry along Boston's Route 128, which became the center of the computing universe. Along with DEC, all the leading minicomputer manufacturers were there, including Prime, Data General, Apollo, and Wang.

For manufacturing, today's fab labs match the cost and complexity of a PDP for computing, again filling a room and initially costing around a hundred thousand dollars. Just as the PDP was a system that together does the job of what we now call a computer, the fab lab can be viewed as a system to do digital fabrication. Fab labs can turn data into things and things into data, but they do so within a lab that entails learning how to use multiple machines. But because fab labs, like PDPs, can belong to a workgroup rather than a large organization, they're likewise being used to explore how digital fabrication will be used when everyone has access.

The first true personal computer was the Altair 8800. Designed by technicians from the Kirtland Air Force Base in Albuquerque for the Micro Instrumentation and Telemetry Systems company (MITS, an acronym intentionally chosen to sound similar to MIT), the computer was launched to make instrumentation kits for model rocket enthusiasts. When this hobbyist computer kit appeared on the cover of *Popular Electronics* in 1975, it was a life-changing experience for the magazine's readers (including me—I still remember sitting in the back seat of my parents' car, clutching the new issue with my jaw hanging open at the implications).

When the Altair originally shipped, the only way to load a program into it was to flip switches on the front panel with the data, and the only way to get results was to watch the front panel lights blink. But because the cost came down to around a thousand dollars, this was a computer that an individual could own. Bill Gates and Paul Allen used a PDP at Harvard to develop Microsoft's first product (or Micro-Soft, as they originally wrote it), a BASIC language interpreter for the Altair. The arrival of the Altair was also the impetus for the first meeting of the Homebrew Computer Club in Menlo Park, inspiring Steve Wozniak and Steve Jobs to launch Apple Computer.

While all that was happening, the Altair was being ignored by the minicomputer industry. Ken Olsen famously said in 1977 that "there is no reason for any individual to have a computer in his home." Although his comment has been widely misinterpreted—he was referring to a computer literally in the construction of the home rather than a personal computer used by its occupants—he still had it wrong, failing to anticipate the rapid spread of smart devices throughout the home. All the minicomputer companies failed when personal computers arrived, because the companies had seen them as uncompetitive toys. As I noted in Chapter 1, DEC was eventually sold to Compaq, which in turn was merged into HP. In retrospect, the organizational-change lesson here is that the minicomputer industry was doomed; the best thing it could have done would have been to give up rather than hang on.

Digital fabrication today is passing through a stage that can be compared to Altair's history. A number of fab lab projects are doing rapid prototyping of rapid-prototyping machines. These are versions of fab lab machines that can be made with other fab lab machines. And like the minicomputer industry's response to the Altair, the consistent response from the manufacturing industry to fab labs has been that these toys might be good for fun things like education or entertainment, but they won't affect the serious business of manufacturing. But these tools can already locally produce many of the products that consumers now purchase from global supply chains. Just as PCs led to the demise of the minicomputer industry, fab labs are likely to lead to new jobs that don't come back to old factories.

The PC emerged in its modern form with the Apple II in 1977 and then the IBM PC in 1981. The PC appeared to be a single unit with an on-off switch, but internally it integrated the roomful of subsystems found in a PDP. There's nothing yet analogous to a PC for personal fabrication. It's not a 3D printer, much as the hype around them might suggest, because these printers are just one of the many digital fabrication machines needed in a fab lab to produce all but the simplest finished products. And it's not going to be the shrinking of all the machines now in a fab lab into a box, because that would still rely on the substantial inventory of the raw materials and waste disposal that these now require. Rather, we'll see that a true personal fabricator, like the universal *Star Trek* replicator, rests on the much more fundamental sense of digital fabrication in coding the construction of materials.

Ⓝ

FOUR BILLION YEARS OF DIGITIZATION

Digital is a good candidate for the word that is simultaneously most widely used and widely misunderstood. The meaning of digital in communication, computation, and fabrication has a common origin that is much deeper than the use of ones and zeros.

Compare the properties of one of my favorite fabrication systems, a child playing with Lego bricks, with one of my least favorite ones, today's 3D printers:

- *Reliability:* Because of the error correction that comes from snapping the bricks together, the Lego construction is more accurate than the motor control of the child. The 3D printer, on the other hand, is only as good as the position measurement of the printing head and will fail if there's a disruption in the flow of material through the head.
- *Modularity:* Lego bricks made out of dissimilar materials can be joined by a standard interface. But 3D printing requires materials that can all go through the same deposition process.
- *Locality:* The child doesn't need a ruler to place the bricks; the global geometry comes from the local parts. This means a child can make something bigger than himself or herself. The 3D printer is restricted to its bed size.
- *Reversibility:* Lego bricks, unlike 3D prints, don't end up in the trash. Trash itself is an analog concept, meaning there's no information in the material to guide its disassembly.

These four attributes—reliability, modularity, locality, and reversibility—are the essential attributes of digital systems for communication, computation, and fabrication that converge in the third digital revolution. We'll look at the digitization of each of those in turn.

In 1931, Vannevar Bush made one of the last great analog computers at MIT: the differential analyzer. It was a room full of gears and pulleys that could be configured to solve engineering equations. The longer it ran, the more the answer diverged from its correct value.

The digital computers that you use today (hopefully) don't do that. You can thank the mathematician John von Neumann, who showed how to compute reliably with unreliable devices. In "Probabilistic Logics and the Synthesis of Reliable Organisms from Unreliable Components," a

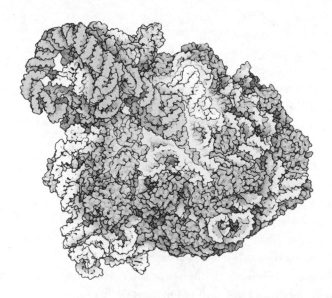

Molecular model of a ribosome in action. *David Goodsell, RCSB Protein Data Bank*

monograph based on a series of lectures he gave in 1952 on work he had done a few years earlier, he proves what we now call a *threshold theorem*. He showed that by computing with discrete symbols (like 0 and 1) rather than continuous quantities (like the rotation of gears and pulleys), if the probability that a device in the computer will make a mistake in an operation is above a threshold (which depends on the system design), then a computation is doomed to fail. But if the probability of a mistake is below the threshold, then the likelihood that the computation fails falls off exponentially as a function of how many times each operation is replicated to check the result. The likelihood of the mistake's happening vanishes so fast that a digital computation can be designed to effectively never encounter an error. They do occur, but very rarely. In all of engineering, there are very few insights that lead to exponential improvements in performance; this is the most important one, and is the real meaning of digital in computation, communication, and fabrication.

Von Neumann built on the work of Claude Shannon, who he met in 1940 at the Institute for Advanced Study in Princeton, New Jersey. In 1937, Shannon had written what is probably one of the most influential master's theses ever, at MIT. After working for Bush on the differential analyzer, Shannon realized that the electromechanical relays used to switch telephone calls could be connected to evaluate arbitrary expressions of true and false statements, called Boolean logic. He didn't just do this abstractly; he demonstrated how to go about designing these universal logic circuits, which were the foundation of all of the computers to follow.

After MIT, Shannon moved to Bell Labs, where ever-greater demands were being placed on the phone system. Rather than make incremental improvements, he answered a profound question that hadn't been asked: what are the fundamental limits? In "A Mathematical Theory of Communication," published in 1948, he developed a theory of information. Along with showing how messages can be compressed, Shannon introduced a threshold theorem for communication. If a message is sent in symbols rather than the continuous analog quantities in use at the time, again if the electrical noise in the phone line is above a threshold then the message is sure to be garbled. But if the noise is below a threshold, then the probability of a mistake in receiving the message falls off as an exponential function of how many times the result is checked. What matters is not the specific values of 0 and 1; it's the use of discrete values that allow errors to be detected. Many digital systems, such as a cell phone, use other sets of symbols that can be more efficient. You're currently receiving a set of twenty-six symbols that can be used to detect ~~errurs~~ errors.

It then took about a decade for the phone system to begin the transition from analog to digital—a transition that led to the Internet. In a sobering lesson about organizational change, Bob Lucky, a thoughtful Bell Labs manager whose employment spanned from Shannon's day to when I worked there, explained that there had been a battle between the analog and digital camps. It was resolved not by persuasion but by attrition: the managers who were analog advocates died out, and a new generation of digital managers took over.

Shannon's Bell Labs colleague John Tukey coined the term *bit* as a contraction of "binary digit" to refer to the smallest unit of information. But the concept of a bit has a physical origin, first appearing in 1929, in Leo Szilard's analysis of Maxwell's demon. In this thought experiment that the physicist James Clerk Maxwell posed in 1867, a microscopic demon could open and close a partition between two chambers and selectively separate fast and slow gas molecules to power an engine indefinitely without appearing to have done any work. Szilard (who would go on to be a leader in both creating and then controlling atomic weapons), reduced this puzzle to its essence with a single molecule and the bit of information being which side of the partition the molecule is on. The connection between information and physics was completed at IBM in

1971 when Rolf Landauer finally exorcised the demon by explaining that it isn't a perpetual-motion machine once the mind of the demon is included in the accounting. He and his colleague Charles Bennett went on to show that abstract thought necessarily consumes physical resources. Landauer and Bennett were my mentors in the study of the physics of computation, introducing me to the concept that information is physical. The inevitable consequence of that observation (for me) was the connection between computation and fabrication.

The link from digitization to fabrication was first made a bit earlier, about four billion years ago. That's the evolutionary age of my most favorite manufacturing machine, the ribosome. This is a molecule that makes molecules. It reads a code, the genetic code, that arrives in a messenger RNA molecule. The genetic code has all the properties of a digital code that Shannon and von Neumann introduced—errors can be detected and corrected, and it's designed so that when errors do occur, adjacent code words produce molecules that have similar properties. But these molecular messages from the genome don't just describe shapes—they *become* shapes.

The symbols in this code are written as A, C, G, and U, which represent the bases adenine, cytosine, guanine, and uracil. Codons, or groups of three of these letters, are matched to one of twenty standard amino acids, which are a bit like molecular Lego bricks. The amino acids are first formed as linear structures in a sequence, which then fold into a secondary structure of geometrical motifs like helices and sheets, which form the tertiary structure of 3D protein shapes. Out of these tertiary structures are built the quaternary structure of functional systems, such as the sensors our bodies use to detect smells or light and the motors that move our muscles. What's remarkable about these amino acids is how unremarkable they are. They have a range of properties, such as attracting or repelling water, or being more acidic or less so. None of these are extreme properties; the chemical behavior is ordinary. Yet through the combination of these standard twenty amino acids, it's possible to make you.

If you mix two chemicals that have a reaction, a yield of 99 percent is considered good, meaning that 1 percent didn't react. The ribosome makes an error once in 10^4 steps when it makes a protein, because it's constructing with a code. DNA replication, which adds an extra error-correction step, has an error rate of one in 10^8. That's the exponential scaling of threshold theorems, and it's what makes possible the complexity of you. The secret of life is that it's digital.

The genetic code carries a message that performs a computation to program fabrication. The third digital revolution has been with us from the dawn of life, but it has been restricted to what can be made with the materials of molecular biology. What has been four billion years in the making is the extension of that insight to the (presently) inanimate part of the world.

INTELLIGENT DESIGN

If the secret of life is that it's digital, then how is it programmed? This isn't a rhetorical question; my lab is part of a collaboration that, led by the J. Craig Venter Institute, is creating living cells by designing them in a computer and then synthesizing their complete genomes. To design a self-reproducing assembler, we'll need to learn how to do the same in systems that we create from scratch, engineering their evolution.

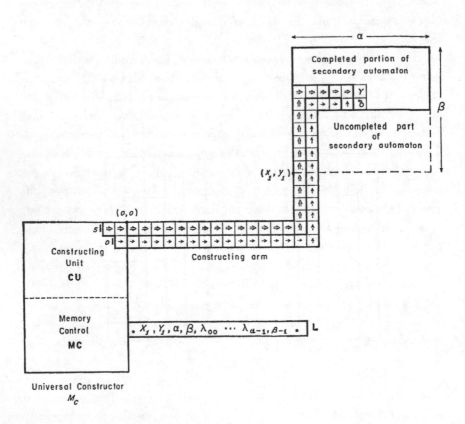

Von Neumann's universal constructor. *Arthur Burks*

The obvious answer to how life is programmed is the genetic code, but this answer is as informative as observing that Shakespeare's plays were written with an alphabet of twenty-six letters. Knowing what the symbols are is only the starting point in learning the language. It turns out that this question is at the heart of the historical alignment of all three digital revolutions.

Making anything with a computer today is like a bad version of the party game where a message gets passed until it is hopelessly corrupted. It starts with CAD, learning how to use computer programs that have had notoriously steep learning curves to design what you want to make. Then comes CAM, converting the design into steps for a machine to follow to make it. That's followed by machine control, converting the CAM instructions into operations for a particular machine. Finally comes motion control, interpreting those operations for the particular parts of a machine. All these steps usually go in only one direction; there's no way for the machine to talk back to the design about problems that arise in making it. And all this has to be redone if anything changes in the design or the machine.

This state of affairs came about because, historically, each of these steps was done by a different person at a different place and time. You could attribute this division of labor to a mistake made back in the Renaissance. That's when the liberal arts emerged as mastery of the means of expression, which meant the trivium and quadrivium (roughly, language and natural science, respectively). When making things was relegated to an illiberal art pursued for commercial gain, art and artisans diverged. This divide is why I was told in high school that I wasn't allowed to go to vocational school, because I was too smart, and why at Bell Labs I was reprimanded for going into the workshop to make something rather than telling the workers there what I wanted them to make. This no longer makes sense if one person can do all these steps at the same place and time. The means of expression have changed since the Renaissance; 3D machining or microcontroller programming can be every bit as expressive as painting a painting or writing a sonnet. Personal rather than mass production demands a new notion of literacy that embraces making as a skill that's every bit as fundamental as reading and writing.

Recognizing that there is a design literacy for digital fabrication will correct this historical error. But it still assumes that there is a designer. Although this observation might sound somewhere between tautological and theological, who designed you? The answer lies in an even deeper connection between computation and fabrication. To understand that, we have to understand how the digital and physical worlds diverged.

I'm happy to take credit for the observation that computer science was the worst thing to happen to either computers or science. By that I mean that the canon of computer science is founded on nonphysical assumptions. It teaches how to write software in languages that are divorced from physical reality and that then rely on someone somewhere else to translate the virtual into the physical. It's a bit like the movie *Metropolis*, where the rulers frolicked in sunlit gardens while, deep below, workers moved the levers that made the city work. And like *Metropolis*, there's a revolution brewing down below.

One of those basement dwellers is my former student Jason Taylor, who as the vice president of infrastructure at Facebook is responsible for building its enormous computing facilities. Another basement dweller is my former student Raffi Krikorian, who had the same job at Twitter. Neither of them studied computer science with me, but on the epic scale that they're working on, you can't believe in software as an abstraction removed from physics—their job is to convert dollars of investment, pounds of equipment, and watts of electricity into information most effectively.

The prevailing segregation of hardware and software is embodied in what's called the von Neumann computer architecture. Von Neumann wrote beautifully and carefully about his seminal contributions, like how to compute reliably with unreliable devices. This wasn't one of those. It comes from a memo that he wrote in 1945: "First Draft of a Report on the EDVAC." As the title suggests, this was not a document for the ages. The EDVAC was one of the earliest electronically programmable digital computers; the report presented his thoughts on how it would be programmed. He suggested dividing it into what he called "organs," like a processing unit and a memory unit. This computer could speed through calculations at a thousand operations per second; any self-respecting smartphone or PC can now do a billion operations per second, but we're still living with the legacy of the EDVAC's limitations in their design.

Von Neumann (and Claude Shannon) had met Alan Turing in Princeton at the Institute for Advanced Study during the World War II code-breaking effort. In 1936, Turing, from Cambridge, England, introduced a theoretical model of computation that was based on a tape containing symbols that are read and written as it moves past a programming head. He used this model, which came to be called a Turing machine, to introduce the notion of a universal computer that can simulate any other kind of universal computer, and then he showed that there are problems that can't be solved by a computer. These profound insights were never meant to suggest a serious computer architecture. Unfortunately, this is effectively what happened, and the parts of Turing's machine became enshrined in the organs in von Neumann's architecture.

The problem with this legacy is that when the head is considered a separate entity from the tape, most of the computer's power is wasted shuttling information around. Very little power goes to actual logical work, even though the memory transistors are as computationally capable as the processor transistors are. And many of the security vulnerabilities in a computer arise from things that are supposed to be logically separated but end up being physically adjacent.

Von Neumann understood these problems. Other than the EDVAC draft report, he never wrote about "his" architecture. The last thing he wrote, which is in the beautiful-and-profound category, is his "Theory of Self-Reproducing Automata," published posthumously in 1966. He was ultimately interested in understanding life. To study this, he co-developed with colleague Stan Ulam a very different model for computation. Called *cellular automata*, the model explicitly represents space and time as well as logic. You can think of it as a self-playing game of checkers, with tokens moving on a regular grid. This was computation in the real rather than the virtual world. It could compute anything that any other universal computer could do, but now operations could happen in parallel everywhere all the time. Von Neumann used this computational universe to design a machine that can reproduce itself, including the instructions for how to do that. This was a theoretical construct at the time, but we'll see in Chapter 5 that a self-reproducing machine is the destination of the research roadmap for the third digital revolution.

Turing also appreciated the physical embodiment of computation. The last thing he wrote (beautifully) about is pattern formation in nature, in "The Chemical Basis of Morphogenesis," in 1952. His goal was to explain how to go from genes to anatomical structure, that is, from bits to atoms. He did this by showing how structure can emerge from the equations describing the spatial distributions of chemicals.

Both Turing and von Neumann were approaching what author Douglas Adams called the "life, the universe and everything" question. The most interesting thing that we humans have evolved is arguably interest itself— our self-awareness. The workings of this evolutionary process resides in a surprising connection between natural and artificial intelligence.

(Marvin Minsky, considered the father of AI, helped plan CBA around that relationship.)

If you lose your keys in a room, you can search the room to find them. If you're not sure which room you lost them in, you have to search all the rooms. If you're not sure which building, you have to search all the rooms in all the buildings. If you're not sure which city, you have to search all the rooms in all the buildings in all the cities. The further along this sequence you go, the more likely the search will be hopeless, until it's nearly certain that you won't find your keys. The technical term for this state of affairs is the *curse of dimensionality*.

What's behind the recent rapid advances in AI is not a breakthrough in understanding intelligence; it's progress in coping with the curse of dimensionality. A conversational computer must identify what you say out of all the possible things anyone can say; a self-driving car must recognize anything that can happen on a road. In what are called *deep learning neural networks*, the mathematical approaches to searching for solutions are those that have been used all along; what's new is having enough data to train the network to make these kinds of predictions, and enough computing power to process the data to enable the layers of the networks to build up good representations of where to search. Something very similar happens in our brain, as sensory input is processed in higher- and higher-level abstractions.

In evolution, the curse of dimensionality is the search for a survival fitness advantage by varying all the possible adaptations of an organism. The way life solves this problem is with Hox genes, one of the most conserved parts of the genome in all living creatures; these genes haven't changed in hundreds of millions of years. Hox genes regulate the expression of other genes. They choreograph what are called developmental programs, which are the steps in going from a single cell to a complete creature. Nothing in your genome says you have five fingers; instead there's a program that, when followed, produces five fingers. It's a program in exactly the modern sense of a computer program, with logic connecting inputs and outputs.

The shorthand way to refer to the connection between evolution and development is *evo-devo*. Evo-devo passes through what's called a phylotypic bottleneck, or hourglass. In the phylum that we belong to, Chordata, there are many approaches to reproduction and many body plans. But once an egg in the phylum Chordata is fertilized, all the embryos appear to be similar before they specialize into whatever they're going to become. This similarity of early life stages was once mistakenly believed to be because "ontogeny recapitulates phylogeny," a once-popular comment meaning that the embryo itself passes through all the stages of evolution. It turns out that's not true. The embryos are all passing through a common solution to morphogenesis, the birth of form.

The human genome has billions of bases, and you have trillions of cells. There isn't enough time in the age of the universe to try varying them one at a time, and even if there were, most of those variations would either be inconsequential or fatal. The Hox genes provide a constraint to a much smaller space to search for alternative body plans among ones that are likely to be viable.

This, then, is the other secret of life: the complexity is a consequence of detecting and correcting errors in the assembly of a small set of discrete building blocks (the amino acids), and the diversity is a consequence of designs being represented as developmental programs rather than construction plans (the Hox genes). "I think, therefore I am" is correct, as long as you recognize that the thinking is done by your molecules as well as your brain.

That's what we're now learning how to do in nonbiological systems. I started this section by describing design as it is done today, which is called *imperative design*: you must specify exactly how to make something. I've ended the section by showing that design in evolution is done by searching in a carefully constrained space of developmental programs. The goal of evolution is survival, but digital design principles aren't restricted to survival. *Declarative design* or *generative design* is the name for a design process that lets you describe what you want something to do, but not how it should do it. For example, you would specify the speed, load, and range of an airplane, but not its size and shape. Early attempts at declarative design were as limited as were early attempts at AI, but progress is now being made in both for the same reason: we are mastering the curse of dimensionality. The search part of declarative design is currently done offline in simulation in a computer before anything is fabricated, but as computation and fabrication converge, design and fabrication will be able to occur online continuously—the airplane itself could evolve in response to changing loads and aerodynamic regimes. At that point, the distinction between animate and automata will become increasingly semantic.

(N)

ALAN & JOEL

CHAPTER 4
The Social Science

When the distinction between animate and automata becomes blurred, the social sciences must become more focused. With digital technologies, one revolution has built on another; the past is a clear prologue to the future. With the social sciences, things are a bit messier. Much can be learned from the past, but the past is by no means a clear prologue for understanding and shaping the future.

Science drove advances in digital communication and computation, just as it is now driving advances in digital fabrication. The social sciences should be equal drivers when it comes to the human impacts of these transformational technologies. Currently that is not the case. In fact, it will take a culture change in the social sciences if they are to play this much-needed role.

The core challenge with the third digital revolution, like the first two digital revolutions, is that technology is advancing at exponential rates while people tend to change at a linear rate. Historically, the ability of individuals and organizations to adapt to accelerating technological change has varied, with the slowest rates of change limiting the full potential of the technology. At times, the technology itself is limiting, but invariably, as technology progresses, social systems become the limiting factors. In this chapter on social science, following a chapter on science, we show how social systems have struggled historically to keep pace with accelerating technology and highlight the importance of different rates of change for individuals, organizations, and institutions. Finally we look at some of the bright spots where methods and mind-sets have emerged and can be leveraged to help shape the third digital revolution.

In the last chapter, Neil described how the battle between analog and digital camps at Bell Labs was resolved not by persuasion but by attrition: the managers who were analog advocates died out, and a new generation of digital managers took over. Or, as the physicist Max Planck observed, a new idea only sees the light of day "because its opponents eventually die." In a world of accelerating change, however, this strategy will not work. We cannot wait a generation for the social systems to co-evolve with the technology.

MOORE'S LAW VERSUS LASS' LAW

To illustrate the importance of ensuring that the social sciences proactively co-evolve with technology, we begin as Neil did in Chapter 3, with Moore's Law. Moore's Law is popularly understood as the driving force behind digital technologies' becoming better, faster, and cheaper. Over the past half century, periodic reports of the end of Moore's Law have consistently been proven wrong, giving it an air of inevitability.

And yet, Moore's Law is as much the product of social forces. It begins, of course, with Gordon Moore's observations. Before it had a name, it was just a published article outlining observations on rates of change and the implications. The article motivated research by the scientists at Fairchild Semiconductors and other pioneering companies. It spurred Moore and Robert Noyce in 1968, when they launched NM Electronics, the predecessor of Intel Corporation. At Intel, the successive doubling in performance became the core benchmark for the corporation and, over time, for its partners, customers, and competitors. As these doubling capabilities improved the performance of countless goods and services, Moore's Law became integrated into social expectations. Maintaining the law became a must-have not only for Intel, but ultimately for society, which enthusiastically embraced ever-better, ever-faster, and ever-cheaper digital devices and applications.

In the social sciences, Moore's Law is what is called a *socially constructed phenomenon*—it is not a natural law, but rather the product of repeated human behaviors. In this case, the competitive environment is socially constructed. In management science, this kind of socially constructed environment is what is termed an *enacted environment*. That is, Moore's Law is a case of business creating or enacting the very environment in which it competes. The continued acceleration of information technology performance happened because people made it a must-have—first in industry and then in society. What would it take for that to happen with digital fabrication performance? To answer this question, we need to understand some key social differences between Moore's Law and Lass' Law, starting with the role of Intel.

Intel Corporation has been ground zero for sustaining Moore's Law for a half century. "Moore's Law *is* our business strategy," says Peter Levin, a social scientist supporting strategic planning in Intel's Data Center Group. "Maintaining Moore's Law has defined business success for Intel. It has offered a set of criteria enabling partners and customers to evaluate the company's engineering and manufacturing prowess over a long period of time." And, he adds, "It hasn't been easy. The cost and complexity of maintaining this competitive advantage is incredibly hard—it takes a sustained financial, research and development commitment on a vast scale. But it is one thing that Intel does better than anyone." He concludes with an evocative metaphor: "Ensuring the continuation of Moore's Law is a bit like Indiana Jones staying ahead of the boulder as it careens down on him in the movie." For Intel, the challenge of maintaining Moore's Law is an ever-present existential threat. Addressing this threat has required sustained urgency, focus, and strong execution.

A key question, then, is for whom is Lass' Law—maintaining the accelerating pace of digital fabrication performance—an existential driver? Although a single driven company could emerge, we believe it is most likely to be an ecosystem of interdependent but independent stakeholders who will be driving the accelerating gains in digital fabrication performance. In that case, it is an open question whether an ecosystem of organizations (public, private, and nonprofit) can have the same existential drive as a single multinational corporation (with executives, employees, and shareholders driving continuously improving financial results). Moreover, such a distributed ecosystem might or might not emerge with the financial and human resources to maintain the needed basic research and development.

Shaping social systems to align around Lass' Law will be one of the defining challenges for society and, hence, the social sciences in the third digital revolution. Doing so in a way that attends to the threshold challenges of fab access, fab literacy, enabling ecosystems, and risk mitigation will be the measure of success. Unfortunately, the track record for the social sciences taking such a proactive role is not great.

For many years after the identification of Moore's Law, nearly all the social science disciplines failed to recognize that the world was on the cusp of massive transformations driven by digital technologies. Because they did not engage deeply with the emerging digital technologies, they did not understand the exponential potential and were consequently limited in their ability to shape the social systems to co-evolve. As a result, we are still playing catch-up.

Reactive critical analysis of technology—that is, observation—is the norm in the social sciences. This approach is necessary, but not sufficient in a world of exponentially accelerating technologies. In the social sciences, there is a need for a mix of path observers *and* path creators when it comes to people and technology. To understand how this might be achieved, we go back to the industrial revolution and the origins of the modern social sciences.

REACTIVE VERSUS PROACTIVE SOCIAL SCIENCE

The social sciences, the study of human societies and relationships, have largely been on a reactive path with respect to technology since the beginning of the industrial revolution. This reactive mind-set has so permeated the social sciences—sociology, economics, political science, management science, industrial relations, social work, and other related fields—that it will be difficult for social scientists to anticipate and help create the path for accelerating technological change.

The industrial revolution was marked by a massive qualitative shift from the preceding craft era—a shift as consequential as the shifts associated with the more recent two digital revolutions. The industrial revolution was powered by a variety of interrelated technological developments across engineering, chemistry, metallurgy, and, crucially, steam power. Steam power, which commercially emerged in the late 1700s, enabled a thousand-fold increase in human power, driving widespread transformations throughout society. The changes included the transition from handcrafted production to industrial mass production and the shift from agrarian and cottage-based economies to factory-driven mass markets. Also relevant was the introduction of new chemical, iron, and textile manufacturing processes accelerating industrial output, further enabled by the development of canals and, thus, large-scale transportation of goods.

With the new technologies came something new—class mobility. Entrepreneurial individuals rose to become owners of the new enterprises, giving rise to what would come to be called the American Dream (hard work and education meant that your birth was not your destiny). At the same time, for the vast majority of the workforce, the industrial revolution brought a Faustian bargain in which steady wages came at the expense of deplorable and unsafe working conditions, oppressive supervision, and environmental degradation.

Most modern social sciences emerged or accelerated in response to the first industrial revolution—many decades after the development of the enabling technologies. Modern sociology began, in large part, in the late 1800s with the connection Max Weber made between religious and cultural principles and the logic of capitalism and with Émile Durkheim's documentation of the emerging division of labor. Social work emerged as a field as both Octavia Hill in England (in the late 1800s) and Jane Addams in the United States (in the early 1900s) responded to the people, particularly women and children, who were being injured and displaced by the industrial revolution. Psychology predated the industrial revolution but took on its modern form with the formation of a professional society in France in 1885 and in the United States in 1892. Industrial relations emerged as a field in England in the late 1800s with Sidney and Beatrice Webb's documentation of oppressive working conditions and the need for new forms of industrial democracy. In the United States, the origins of the field in the early 1900s centered on John R. Commons's findings that institutional arrangements, such as the shift from craft unions to industrial unions, followed changes in markets and technology. In these and other cases, because the founding scholars for the social sciences were reacting well after the impacts of the industrial revolution had become deeply entrenched in society, the social sciences were constrained in their capacity to improve lives, mitigate harm, and deal with the entrenched interests that were opposed to social change.

The industrial revolution also inspired new themes in fiction, reflecting the impcts of technology on society. Mary Shelley's *Frankenstein*, released in 1817, was an exploration of the good and evil faces of technology. This is evident in this passage from the point of view of the monster, where fire can be understood as a metaphor for technology: "One day, when I was oppressed by cold, I found a fire which had been left by some wandering beggars, and was overcome with delight at the warmth I experienced from it. In my joy I thrust my hand into the live embers, but quickly drew it out again with a cry of pain. How strange, I thought, that the same cause should produce such opposite effects!"

As the full impact of the industrial revolution became evident, the observations by the storytellers more often focused in on the harm associated with the technology. Charles Dickens's 1854 novel *Hard Times* took aim at policies that reinforced the oppressive conditions in factories. Similarly, Upton Sinclair's *The Jungle* in 1906 horrified the public with its images of unsafe working conditions, child labor, and sexual harassment in the meatpacking industry. These and many other works of fiction reinforced the efforts of social scientists on public policy even if the system was already deeply entrenched.

Today, the founding scholars and socially aware writers who documented the impacts of technology on society are correctly honored as pioneers—defining new fields of study and bodies of literature. Yet their approach to the technology of the industrial revolution was centered on observing and addressing its impacts after the fact. Child labor laws and early experiments with labor-management cooperation that emerged could have been far more effective and could have alleviated much more suffering if they had been given full consideration fifty or a hundred years earlier, as the technologies were emerging.

Addressing technology after it is well established limits what is possible moving forward. As organizations are established, a phenomenon known as *path dependency* is set in motion in which past decisions under a given set of circumstances constrain future decisions, even if circumstances are different. As a result, once on a given path, organizations have a strong tendency to stay on that path. The concept of path dependency comes from the social sciences, and, ironically, the social sciences themselves have been largely captive to a path with respect to technology. On this path, social sciences are reactive rather than proactive. To a great extent, the social sciences are still constrained to this path.

Reinforcing path dependency is another dynamic, which is what Robert Michels in 1911 labeled *the iron law of oligarchy*. He observed that new institutional arrangements (he focused on political parties and trade unions) come into existence because of a given mission, but then give priority to their continued existence, rather than whatever the initial mission was. These same dynamics were documented with respect to scholarly fields and disciplines in 1962 by Thomas Kuhn, who wrote *The Structure of Scientific Revolutions*. Kuhn observed that it took revolutionary change to overturn established institutional orders in science.

We call the reactive approach to the study of new technology *path observation*. With this term, we have two meanings of *observation* in mind—the reactive approach is more one of observation rather than shaping the technology, and it involves observing, or complying with, the established rules and norms, reinforcing the path observation approach. This approach is strongly enforced in the social sciences through peer review and other mechanisms. Path observation, of course, is necessary, but not sufficient if the social sciences are to take a leadership role in helping to shape a world of accelerating technologies.

Proactively helping to shape the third digital revolution will require nothing less than a culture change in the social sciences. The dominant "positive" social science approach, which is one of path observation, will need to expand to include a much more extensive application of the "normative" approach. Positive social science focuses on explaining what has happened. A normative approach includes views on what should happen. It still must be grounded in evidence, including a clear understanding of the underlying science and the technology. By itself, normative work is often disparaged as not being science—though there are dedicated subcultures committed to what is variously termed action research, participant observation, and similar approaches. It is possible, indeed essential, to combine the positive and normative approaches by embracing and advancing bold, evidence-based analysis of accelerating technological developments and desired transformations.

The third digital revolution provides social science with a rare opportunity to engage with emerging new technologies while they are still in formation. It is a chance to look around the corner into the future implications across the many fields and disciplines—economics, sociology, political science, management science, industrial relations, and others. This will require the social sciences to shift into a more proactive stance to become path creators, not just path observers.

Path Creators

Among social scientists, Adam Smith developed early principles of economics in the middle 1700s, very close to the beginning of the industrial revolution. In this case, his concept of the invisible hand (i.e., self-interest), the division of labor, and free markets became interwoven with the technologies of the industrial revolution. His impact illustrates the power of being proactive early in the development of new technology. Crucially, he came to these insights through a careful examination of the technology itself, as evidenced by his detailed description of the division of labor and use of technology in a pin factory. He was a path creator or, more accurately, a path co-creator with the entrepreneurs and emerging industrialists of the time.

By the time Karl Marx and Friedrich Engels came along in the mid-1800s, the technology of the industrial revolution was already well advanced. They focused on the control of the means of production. The ownership class was identified as the root cause of the problems in society since, as Marx analyzed, they were just extracting value from the work of others and keeping it for themselves in the form of profits. Since the concept of ownership of capital was, by that time, deeply embedded in society, the only path that Marx and Engels could envision was that of a revolutionary overthrow of the ownership class and replacing it with a form of social ownership. In this sense, they were path creators, inspiring revolutions worldwide, centered on a set of operating assumptions contrary to those driving the industrial revolution. With Marx and Engels, we see that the work of path creation was progressively more difficult the longer it happened after the technology emerged and as technological assumptions became embedded in social systems. The third digital revolution provides a new opportunity to reshape the system. Digital fabrication, with its distributed control over the means of production, can fundamentally change the nature of work—reducing the extractive properties of the system born during the industrial revolution, without requiring a revolutionary overthrow of that system.

There have been many notable path observers since Marx and Engels, yet very few have aspired to be path creators with respect to technology. Frederick Taylor stands out as a path creator, but his innovations actually had the effect of making people subservient to the technology. Beginning in 1911, with the publication of *The Principles of Scientific Management*, Taylor pointed to "the great loss which the whole country is suffering through inefficiency in almost all of our daily acts." He argued that "the remedy for this inefficiency lies in systematic management, rather than in searching for some unusual or extraordinary man." The ideas took hold—Taylor Society chapters sprung up around the nation. Through the time-and-motion study of work and other means, Taylor did show how to increase efficiency by, in effect, treating people as additional cogs in the machine. The resulting field of industrial engineering emerged from this work, combining the social and the technical, but too often with the social as handmaiden to the technical.

After World War II, some social scientists tried to promote an alternative path with respect to people and technology. Under the broad umbrella of what came to be known as the postindustrial movement, they took on what we are calling path creation in their efforts to move beyond Taylor's mechanistic logic and reshape technology toward a more humanizing path. Daniel Bell, a leader in this movement, wrote in 1973 that technology "is a form of art that bridges culture and social structure, and, in the process, reshapes both." His warning that "technology is always disruptive of traditional social forms and creates a crisis for culture" is equally true today as we consider the third digital revolution. However, he and others in the postindustrial movement missed what is arguably the most important of the postindustrial developments: the exponential rate of change of digital technologies.

A key part of the postindustrial movement focused on the integration of the social and the technical in factories and other organizations—advancing our understanding of sociotechnical systems. By going out in organizations and partnering with workers and managers, researchers conducting pilot experiments successfully identified ways to humanize work in diverse technological settings, including British coal mines, Swedish auto factories, American pet-food factories, Danish shipping companies, and Canadian oil refineries. Yet the technology was largely seen as changing incrementally. Even though the sociotechnical systems folks were path creators, not just path observers, they did not create paths with the pioneers of digital technologies. As a result, they were unaware that technologies were on the cusp of exponential change. By failing to position the integration of the social and the technical against the backdrop of what would become transformational technological change, the pilot experiments missed out on a key source of leverage.

Fred and Merrelyn Emery, both part of the sociotechnical systems movement, did address some aspects of the digital technologies as early as 1976, predicting that "the growing alliance between telecommunications and computers" would increasingly "form networks in which the collective information and processing capabilities will be available to all users." Anticipating Neil's claim that the future is present now, they said, "We took the view that we were probably already into our future; but in such a small way that it was not easily recognized." The new technology, they argued, would allow for "the spontaneous generation of content itself, which would make providers and users one and the same." This aspect of digital communications and computation is exactly what we now see characterizing digital fabrication. In 1981, Eric Trist, another early sociotechnical systems leader, built on this analysis: "The oncoming information technologies, especially those concerned with the microprocessor and telecommunication, give immense scope for solving many current problems—if the right value choices can be made." Again, this remark is directly aligned with the aim of this book.

Although these and other critical observations about technology were intended to encourage the co-creation of alternative futures, they missed the exponential capabilities of digital technologies. In their 1977 book *A Choice of Futures*, Emery and Emery were even critical of "Shannon's so-called Theory of Information." Certainly, the core concepts of error correction without degradation in quality were hard to accept. Yet, without an understanding of these essential properties of the technology, path creation was constrained. Imagine what the impact of the sociotechnical systems approach would have been if Emery and Emery had been able to anticipate and ride the wave of exponential growth that was just beginning at the time.

The exponential power of the technology did not go unnoticed by everyone outside the insular world of technology development, but it was mostly appreciated by people out of the mainstream of the social sciences. Stewart Brand, founder of the *Whole Earth Catalog* and *CoEvolution Quarterly*, advocated a decentralized, personal view of technological development—a view that was intended to be liberating and ecologically responsible. Issues of *CoEvolution Quarterly* were devoted to imagining everyday life in space colonies (1975), solar water heaters in Los Angeles (1977), the threats of genetic toxicity (1979), and other topics covering whole systems, shelter and land use, industry and craft, communications, and community.

Although Brand was not a social scientist, his full and early engagement with the technology illustrates what path creation can look like. In fact, he went beyond path creation to a form of co-creation with some of the leading technology pioneers. It led him to the conclusion that co-evolution was not only possible, but also necessary. John Markoff, who wrote *What the Dormouse Said: How the Sixties Counterculture Shaped the Personal Computer Industry*, says: "Stewart was the first one to get it. He was the first person to understand cyberspace. He was the one who coined the term personal computer. And he influenced an entire generation, including an entire generation of technologists." In Steve Jobs's now famous 2005 commencement address at Stanford, he concluded with an extended tribute to Brand and the deeply human values Brand brought to the technological enterprise. Interestingly, Joel, Neil, and Alan all subscribed to and avidly read *CoEvolution Quarterly*, illustrating its appeal across science, social science, and the humanities.

In *Soft-Tech: A Coevolution Book*, which was coedited by Brand and inventor J. Baldwin, the lead chapter, "One Highly-Evolved Toolbox," by Baldwin, describes his "portable shop" that had been evolving for fifteen years. He concludes that "the ultimate is to make your own tools." Brand was a co-creator with a wide range of technology pioneers. His access was both formal and informal. It was a legendary party scene and this was not incidental to the co-creation—high-trust social relationships enabled Brand to bridge the social and the technical in ways that would not be possible through arm's-length observation. Along with inventor Douglas Engelbart, Brand co-delivered "the mother of all demos" at the 1968 annual joint meeting of the Association for Computing Machinery and the Institute of Electrical and Electronics Engineers. The demo featured new technologies such as email, hypertext, the mouse, and others.

Another person who was outside the social science mainstream, but who also tuned into the accelerating pace of change with the technology, was author and futurist Alvin Toffler. In his 1970 book *Future Shock*, he captured the challenge of keeping pace with accelerating technologies and the importance of proactively shaping them:

> Throughout the past, as successive stages of social evolution unfolded, man's awareness followed rather than preceded the event. Because change was slow, he could adapt subconsciously, "organically." Today unconscious adaptation is no longer adequate. Faced with the power to alter the gene, to create new species, to populate the planets or depopulate the earth, man must now assume conscious control of evolution itself. Avoiding future shock as he rides the waves of change, he must master evolution, shaping tomorrow to human need. Instead of rising in revolt against it, he must, from this historic moment on, anticipate and design the future.

Toffler was prescient about the deep psychological impact of too much technological change in too short a time. He argued that we must unleash the "forces of conscious evolution" to shape the impact of the technology—appreciating of the need for co-evolution and designing the future.

Brand and Toffler are to be credited for seeing the importance of path creation in a world where technology is changing at an accelerating rate. Arguably, the limitations of the first and second digital revolutions would have been even worse were it not for the injection these humanistic path creators, but counterfactuals are always hard to confirm. As a result, we see their contributions as exemplary, but like the handful of path creators before them, they were outliers when it came to prioritizing co-evolution of social and technical systems.

The challenge of co-evolving social and technical systems was crystalized in 1984 by Michael Piore, a political economist, and Chuck Sabel, an economic sociologist. Piore and Sabel identified the transformational potential of digital technologies and our entry into what they termed the *second industrial divide*. They understood that computers were combining with manufacturing in new ways that would fundamentally change the nexus between markets and technology—making *flexible specialization* possible. Their analysis begins in the traditional path-observation mode but then moves into path creation as they outline what would be needed for a beneficial future.

Beginning in fashion and publishing, Piore and Sabel document how computers and communication technologies are pointing toward specialized manufacturing that could quickly adapt to niche markets. They also note that "the existing institutions no longer secure a workable match between the production and consumption of goods." They conclude that "these institutions must be supplemented or replaced."

Piore and Sabel were awarded a MacArthur genius grant for their insights in digital technology and the future of work. Their understanding of the flexibility of the technology, particularly around rapidly responsive fashion design and production in the communities of Northern Italy (think of the fashion firm Benetton), anticipates in some ways the emergence of community fabrication. And yet Piore and Sabel's analysis was now two decades after the publication of the Gordon Moore article. The researchers did not fully foresee and account for the potential for exponential growth with digital technologies. While it would have been hard to envision in the mid-1980s, had they understood and fully addressed the accelerating aspect of digital technologies, they would have been more able to help us anticipate the implications for individuals, organizations, and institutions.

Inventors and Entrepreneurs

When it comes to new technologies, the most forward thinking and impactful path creators have typically been inventors and entrepreneurs. They generally operate with a single-minded focus on translating an innovation into a successful commercial enterprise. In the late 1970s and early 1980s, many technologists certainly recognized and fully engaged with the exponential nature of digital technologies. Bill Gates and Paul Allen's early mission statement at Microsoft envisioned "a computer on every desk and in every home" (presumably running Microsoft software). This mission was predicated on an assumption of the exponential nature of digital software. Xerox PARC, Intel Labs, and others engaged social scientists who were exploring the intersection of digital technologies and society. These entrepreneurs and corporate researchers undoubtedly recognized the power and potential of personal computing to have an impact on society, but their primary focus was on building successful businesses, not leveraging the emerging technologies to address societal challenges or mitigate harm. What if the young entrepreneurial Bill Gates had had the mind-set of his elder self, bringing his current social entrepreneurship mind-set to the early formation of Microsoft?

In the 1980s, inventor and futurist Ray Kurzweil began using models to track exponential digital technologies to predict the future. In his book *The Age of Intelligent Machines*, he made a wide variety of predictions based on the accelerating technologies he was observing. But Kurzweil was looking at the models as an inventor (his modeling of future capabilities informed his inventions so they were timed to leverage these emerging capabilities) and as a futurist, not as a social scientist. Through his books, he helped raise awareness of the importance of exponential digital technologies, but he wasn't focused on providing models for how individuals and organizations could co-evolve with these technologies.

In parallel with the inventors and entrepreneurs who tuned into the exponential rates of change were writers of fiction—particularly those writing hard science fiction, that is, material grounded in scientific accuracy—who served as co-creators operating at the interface of people, technology, and the future. These writers were often prescient (and evocative) in painting a picture of the social implications of accelerating digital technologies. In 1968, Arthur C. Clarke's *2001: A Space Odyssey* imagined the power of supercomputers, AI, and even cool gadgets like the "newspad" (an early vision of the iPad). In 1984, William Gibson's *Neuromancer* envisioned the social and cultural impact of "cyberspace," popularizing the term he coined a few years earlier in his short story "Burning Chrome." Of course, Neil has been inspired by *Star Trek* to help turn the replicator from fiction to reality. These science fiction creators (and others like them) not only developed evocative images of future digital technologies, but also uncovered and challenged critical embedded assumptions about these technologies by placing them in a social context with deeply human narratives.

Not surprisingly, technology entrepreneurs and science-grounded storytellers have been more proactive in literally and figuratively shaping emerging technologies than social scientists have been. But it does not need to remain that way. The challenge is to be both positive (in the scientific sense) and normative. The social sciences do not need to abandon their core principles and methods. They need to expand their traditional methods and work across disciplines to understand *and* identify ways to engage and shape technology before it shapes us in ways we will regret.

The social sciences can also be effective for understanding and addressing the ways that technology can shock and disorient people (which is likely to happen with each new phase of the third digital revolution). This begins with social scientists first taking the time to understand the underlying technologies and communities. The humanities can help anchor deeply human values and narratives when we are constructing future scenarios that integrate new technology with social systems. Of course, technologists and scientists will have to be open to inputs and collaboration with the social sciences and the humanities—an attitude that will also require some culture change. For all parties, a closer look at how individuals, organizations, and institutions are able to respond to accelerating change is needed.

RATES OF CHANGE

Digital technologies are capable of exponential rates of change. Social systems and people also change, but typically at more incremental rates. Rates of change vary for individuals (changes in attitudes, behaviors, skill sets, and access), with most changes proceeding in small steps. Change also occurs incrementally for organizations (changes in strategy, structure, process, culture, and technology) and institutions (changes in societal norms, standards, laws, and priorities). Calling for social scientists, storytellers, and others to co-create with technologists requires close attention to these differential rates of change.

Incremental rates of change in social systems serve as rate limiters for the potential of accelerating technologies. Note that we focus on rate limiters more than absolute barriers. There can, of course, be absolute barriers to change, but it is more common that various forms of friction slow down rather than stop change altogether.

When it comes to rates of change, the three domains that constitute social systems—individuals, organizations, and institutions—are analytically distinct, but interwoven in practice. As a result, it is important to understand the rates of change in each domain, as well as how they then fit together with respect to technology. Once the different rates of change are understood, we can develop opportunities for increased influence over how change unfolds. Innovations that have faster rates of change can exploit cracks in slower-moving organizations and institutions, breaking free of rate-limiting forces.

Of course, technology itself can sometimes be the rate limiter. For example, the invention and installation of fiber-optic cable addressed an important technological constraint with the first digital revolution. So did issues like hard drive capacity, hard drive density, and processor speed. Looking ahead, failing to advance the fundamental science of digital materials—a truly end-to-end fabrication process where we can code the construction of the materials themselves—will be a rate limiter for Lass' Law. When technology is the rate limiter, the challenges that need to be addressed focus more on resources for research and fundamental science.

Note that rates of change for social systems vary from no change to linear change to exponential change. We have all experienced no change. This is the frustration of gridlock and the deeper frustration when we realize that some stakeholders see it in their interest to create the gridlock. We have all experienced linear change. We are used to changes in interest rate percentages, productivity rates, unemployment rates, and the like. The rates do fluctuate, and the fluctuation can be consequential, but it is all within ranges that we intuitively understand.

Thinking in terms of exponential change, however, is not instinctual. A number of today's authors use the metaphor of a car traveling at different speeds to illustrate the acceleration of technological change. The metaphor begins with a car going six miles per hour, representing the craft era. Navigating the car is not difficult. A car traveling an order of magnitude faster, at sixty miles per hour, stands for the last century—the industrial era. Starting, turning, and stopping becomes more difficult, but the roads can still be navigated. The postindustrial, digital era is analogous to the car breaking the land speed record, moving at six hundred miles per hour— where the smallest disturbance in the road can be massively disruptive and survival depends on the ability to see further down the road than human eyesight allows. The implication is that individuals, organizations, and institutions will need new capabilities to function effectively when the pace of change accelerates by an order of magnitude.

Staying with the car analogy, continued exponential growth in the rates of change, which is what is projected for digital fabrication, would involve a jump from six hundred to six thousand miles per hour in another decade, followed, midcentury (with accelerating rates of change), by, possibly, six million miles per hour. Hold on tight—it is going to be quite a ride. As Kurzweil cautions, "Today we anticipate continuous technological progress and the social repercussions that follow. But the future will be far more surprising than most people realize, because few observers have truly internalized the implications of the fact that the rate of change itself is accelerating."

Looking back at how institutions, organizations, and individuals navigate change is helpful as we look forward to all three proactively shaping the third digital revolution.

Institutions

Historically, institutions—the human-made laws, regulations, customs, and arrangements that govern what should and should not occur in society—have been slow to change. Begin with the word *institution*, which was coined in French in the middle 1400s. This is the first time it appeared in any Western language. Yet the things it described—the monarchy, the military, the church—had been around for more than a thousand years. The very slow pace of change in these arrangements—measured in millennia—likely accounts for the delay in the word's appearance. For individuals and organizations, there was no need for a word for a concept that wasn't evident across multiple generations.

With the arrival of the industrial revolution, the pace of change accelerated by an order of magnitude. New institutional arrangements—constitutional democracies, political parties, labor unions, corporations, Boys and Girls clubs, Kiwanis clubs—emerged in fifty to one hundred years, rather than five hundred to a thousand years. The pace of change accelerated. Within a lifetime, major transformation was evident and we labeled it a revolution.

In today's digitally powered world, there is pressure for institutions to change at an even faster pace. Institutional arrangements in settings as diverse as music, finance, government, marriage, business, labor, education, security, and religion are all experiencing massive disruption at a pace that is yet another order of magnitude faster than during the industrial revolution. Institutions today must be in a constant state of change, with fundamental shifts happening often—every five to ten years, which is an order of magnitude faster than was the case during most of the twentieth century. And this pace of change continues to accelerate, with some sectors experiencing dramatic shifts measured in months, not years.

The problem is that the institutions are not adjusting anywhere near this quickly. Of course, we need institutions to provide stability and continuity, but they need to balance their core center of gravity or mission with unprecedented foresight, modularity, and agility. If they remain purely reactive and rigid, the results are a lack of relevancy, lost opportunities, stasis, and, often, direct harm to individuals, communities, industries, and even entire societies.

The World Bank's 2016 World Development Report on Digital Dividends points to "analog" aspects of society, particularly institutional arrangements, that are needed to address gaps resulting from the first two digital revolutions: "To get the most out of the digital revolution, countries also need to work on the 'analog complements'—by strengthening regulations that ensure competition among businesses, by adapting workers' skills to the demands of the new economy, and by ensuring that institutions are accountable." What the World Bank termed "analog components" are institutional arrangements that need to operate more like the digital technologies with which they are interacting. There are clues in the very nature of the technology to help accomplish this goal, such as the roles of modularity and error correction in digital technologies, a connection which we develop more fully in Chapter 6. There are also some clear lessons from the more recent history with the first two digital revolutions in how to accelerate rates of change at the institutional level of society.

Consider an illustrative institution: K–12 education. This sector has been remarkably resistant to change through the first two digital revolutions. The slow pace of change for educational institutions has left them vulnerable to disruption. Agile individuals and organizations can exploit the cracks in the system with faster rates of change that are better aligned with the needs of both teachers and students. These emerging institutional arrangements will ultimately be able to succeed, in part, because they leverage and co-evolve with digital technologies in new and innovative ways.

Most of K–12 education is dominated by twentieth-century classrooms (the industrial model), a nineteenth-century structure for the academic calendar (the agrarian model), polarizing battles over teacher and student assessment, and a publishing ecosystem dominated by a handful of entrenched organizations. And yet, growing choruses of education researchers, business leaders, politicians, philanthropists, entrepreneurs, and educational reformers are passionately arguing that we are not effectively cultivating the necessary twenty-first-century skills and mindsets—critical thinking, creativity, collaboration, communication, design thinking, problem solving, resiliency—required for the next generation to survive and thrive in a rapidly changing, complex, and digitally infused world. Learning to learn is key to navigating a world where you are likely to change jobs every two to three years (according the US Bureau of Labor Statistics) and where many of these jobs have not yet been invented. Students may also come of age in a world with fewer and fewer jobs available because of continued advancements in automation, AI, and globalization.

But schools, even those eager for innovation, struggle with the introduction and integration of technology-mediated and project-based learning. To keep pace with the digital age, schools, districts, states, and even entire countries have purchased a lot of technology. Much of this technology, however, lies fallow or is underutilized because teachers aren't trained for tech-mediated learning, the technology and supporting infrastructure (such as Internet connectivity) are unreliable, and there is a dearth of good, pedagogically grounded resources designed to take advantage of the technology. Many parents, who are frustrated that their kids are spending, on average, eight hours a day immersed in digital technologies (at home and in school), are not excited about even more technology in their kids' lives. Administrators who worry about liabilities and safety often lock down Internet access, creating blacklist (blocked) and whitelist (approved) websites. On top of all of these challenges, there continues to rage a political battle around federal versus local control of education. The teachers have been caught in the middle. Often underpaid and always overworked, they find a difficult job even more difficult because of all the institutional friction.

Alan and his business partner Michael Angst encountered this complex institutional ecosystem firsthand when their company, E-Line Media, introduced two innovative game-based learning projects for K–12 schools. The first project, *Gamestar Mechanic*, is a platform that teaches middle school youth how to make games to help cultivate design thinking and twenty-first-century skills. The game, which also provides highly engaging motivation for STEM learning, was originally funded by the MacArthur

Foundation and was developed in partnership with the Institute of Play. The second project, *MinecraftEdu*, was a "mod" (modification) of the popular consumer game *Minecraft*. The mod was originally developed by Teacher Gaming, a company founded by two teachers eager to use the power of the hugely popular game in the classroom. E-Line partnered with Teacher Gaming to help expand the distribution and impact of *Minecraft-Edu*. Together these game-based platforms reached over a million students in more than fifteen thousand schools.

Unlike the textbook publishers with massive sales forces, E-Line has no dedicated salespeople. Its products don't sell into the traditional school channel; they seep in, largely from the bottom up, by going directly to teachers. The rigidity of the existing institutions led E-Line to focus on teacher discovery, teacher-to-teacher recommendations, and hands-on trial in the classroom and at teacher professional conferences. This approach required continually reducing friction for adoption, designing for flexible and transparent pricing, and enabling continual optimization of the service based on teacher and student feedback. Like many other providers of innovative digital-centric learning products and services, E-Line used the power of digital networks and teacher communities to circumvent institutional rate limiters such as centralized purchasing practices, annual decision-making timelines, teach-to-the-test priorities, and fear of the new. Agency was shifted to the teachers, who actually used the products and services, rather than centralized selection and purchasing divisions.

This seep-in strategy took advantage of emerging cracks in the entrenched K–12 system. Although there are some unique challenges for fab-based learning, especially given the complexity of introducing hardware, software, and physical materials, many of the same principles apply when innovative, project-based learning is being introduced to schools. When a growing number of agile, aligned, and passionate organizations take advantage of these cracks, their collective influence can become increasingly deep and pervasive—offering the potential to trigger significant changes across the education ecosystem—enabling new market leaders, platforms, and cultures to emerge. Institutions are the product of repeated, patterned behavior, and these new patterns can become the new institutional arrangements.

Institutions don't usually reinvent themselves. In Chapter 6, we will explore how institutions can make institutions, analogous to machines making machines. History, however, records some powerful forces arrayed against the idea of agile and adaptive institutions. As we noted earlier,

Robert Michels's iron law of oligarchy and Kuhn's scientific revolutions point to *punctuated-equilibrium models*, where periods of stability are punctuated by periods of rapid change. Kuhn documents how the potential for earlier and more continuous change was present, but how rate limiters—in the form of conservative gatekeepers of knowledge—generate increased pressure for change so that, when it happens, it is often rapid and disruptive.

Consider Neil's seemingly simple request for an .edu Internet domain for the Fab Academy from EDUCAUSE, the institutional arrangement that governs these designations, which provides a good example of inflexible rule making within institutions. EDUCAUSE is a new institutional arrangement, born of the second digital revolution, but it still continues to think in some very traditional ways. The rules of the game are that the only educational organizations that can get an .edu designation are those with a physical location. This rule does bring a certain stability to the .edu space, but it also limits innovation and growth for legitimate, nontraditional entrants. The new fab educational model emerged with a faster pace of change than what even a digital institutional arrangement could respond to. The key question is whether EDUCAUSE and other digital gatekeepers can reinvent themselves to match new realities or if they will be supplanted by those that can continually adapt.

Beyond institutions being slow to change, there are also rate limiters associated with the ability of small interest groups to create institutional gridlock. For a window into this dynamic at a community level, consider what emerged when Joel and his team worked with local leaders in Dodge City, Kansas, to do a stakeholder map on digital inclusion. Out of one hundred stakeholders surveyed, ninety-seven expressed varying degrees of support for digital inclusion based on its importance for education, health information, workforce development, and civic engagement. Yet, a few people responded with strongly worded oppositional comments such as this statement: "I think that Internet access is still a privilege, and I am not willing to pay more for my access just so lower-income families and such do not have to pay or get a reduced cost for theirs." This argument that is surfacing is known as *the tragedy of the commons*—the failure of individuals to contribute sufficiently to sustain common resources. At the community summit, considerable attention had to be given to ensuring that these views didn't undercut the entire initiative. Elinor Ostrom earned a Nobel Prize documenting how new institution forms such as public-private partnerships are needed to overcome narrow self-interest (which will continue to be a challenge in the third digital revolution).

The third digital revolution will need institutions that are agile in the face of exponential change. At a functional level, institutions have to do two things, create value and mitigate harm. Or as Henry Kissinger once put it with respect to the institutions of foreign policy: "What is in our interest to prevent?" and "What is in our interest to accomplish?" Institutions still need to advance these functions in an era where driverless cars are making ethical decisions and the ability for almost anyone to hack genes seems to be around the corner, but at a much faster rate of change. Ralph Cicerone called out the need for institutional change on a broad scale in his 2012 presidential address to the National Academy of Sciences:

> Today's most troubling and daunting problems have common features: some of them arise from human numbers and resource exploitation; they require long-term commitments from separate sectors of society and diverse disciplines to solve; simple, unidimensional solutions are unlikely; and failure to solve them can lead to disasters. In some ways, the scales and complexities of our current and future problems are unprecedented, and it is likely that solutions will have to be iterative. . . . Institutions can enable the ideas and energies of individuals to have more impact and to sustain efforts in ways that individuals cannot.

Organizations

The third digital revolution will have a deep impact on all organizations— businesses, nonprofits, government agencies, philanthropies—whether they have roots in an agrarian craft era, the industrial revolution, or the first two digital revolutions. In most cases, the rates of change in strategies, structures, and processes will be faster than those in organizational culture, which is a key rate limiter for organizations. Culture, in turn, is reflected in and sustained by deeply embedded operating assumptions. An organization that survives and thrives in the third digital revolution will need to maintain (and adjust) its operating assumptions so that it is aligned with the new technologies as the innovations emerge. Concurrently, operating assumptions will shape what new technologies can even emerge.

An operating assumption is not usually stated, but it represents the way things happen in an organization. Organizational scholars liken these assumptions to the strands of DNA or code that makes an organization what it is. A foundational operating assumption for organizations was identified by Douglas McGregor in 1960 in *The Human Side of Enterprise*. He contrasts "Theory X" management styles (workers need to be monitored and controlled) with "Theory Y" management styles (work is as natural as play and workers just need to be given the tools and resources to do the best job they can). McGregor observes that "the next time you attend a . . . meeting, . . . tune your ear, to listen for assumptions about human behavior, whether they relate to an individual, a particular group, or people in general." His point is that once you know the underlying operating assumption, you can also predict all that will follow—a set of Theory X assumptions will drive very different decisions and actions than will a set of Theory Y assumptions. In anticipating the implications of the third digital revolution for organizations, we need to learn to tune our ears to the operating assumptions that limit rates of change.

When operating assumptions impede the needed transformation, they must change, and change is not easy. Consider a firm that is emblematic of the first industrial revolution: the Ford Motor Company. Ford has transformed itself from a mass-production behemoth into a more agile, team-based operation demonstrating continuous improvement in quality, safety, and other key measures. (In Europe, mass production is still referred to by social scientists as *Fordism*.) Gonzalo Rey, who was the chief technology officer for Moog, a manufacturer of specialty components for various industries, compares the rate of change over the past century for the shop floor at Ford with other sectors: "If you freeze-frame the work on the shop floor in Henry Ford's factory and look at it over a hundred years, you will see dramatic changes due to technology. Compare that to a construction worksite or a hospital, where the labor content will not have changed much over the same hundred years. The result is that transportation has become widely affordable, while the relative cost of a skyscraper or a home or a hospital visit is not much more accessible than it was a hundred years ago." For Gonzalo, the difference is not just the availability of the technology, but the actual process by which it is adopted and integrated into the organizational functions. That is the key rate of change.

Joel has been deeply involved in the significant changes involving Ford and its US union, the United Automobile Workers, since the 1980s. The key to progress has been the shifting of the company's operating assumptions. For example, managers at Ford have had to move from doing everything possible to contain problems so that the hierarchy would not see them, to openly sharing problems and even discussing near misses to learn from things that might have been problems. This shift in operating assumptions from concealing problems to sharing them required behavioral change at all levels of the organization. At the front lines, workers had to have the confidence that they would not be blamed for reporting a quality error or even stopping the assembly line to contain a problem. Stopping the assembly line costs around ten thousand dollars for each minute it is down.

The same defensive approach ruled at the executive level, where leaders were focused on containing problems and keeping them from becoming a career-ending event. This defensiveness was a rate limiter that turned into a rate accelerator when the culture shifted to support sharing and collectively addressing problems. These changes in the management organization were crucial, but they also took many years to effectuate—the transformation from a culture of blame to one of sharing problems took over a decade and is ongoing.

Similarly, union leaders have evolved from directly opposing unfair decisions and unilateral actions by management to joint partnerships in safety, quality, and other objectives. Union members now earn "black belts" in six sigma change processes and help design quality and safety operating systems. When Joel facilitated a joint quality charter between the United Automobile Workers and Ford, the resulting document was important, but the process was more so—just as the process of revisiting of the fab charter will be at least as important as the resulting document. With the UAW and Ford, the chartering process took a few months, but aligning the work across the enterprise in support of the charter—with operating assumptions centered on partnership and reciprocal responsibility—has taken years. Altogether, the UAW-Ford transformation took thirty years and at least fifty-six pivotal events (like the chartering). In a similar way, it will take many years and more than a few pivotal events for a fab charter to achieve its full potential in bringing people together and changing operating assumptions in the fab ecosystem.

The hierarchical mass-production model is an organizational rate limiter because it concentrates planning, decision making, and supporting actions within the chain of command. Back in 1960, McGregor observed, "It is probable that one day we shall begin to draw organizational charts as a series of linked groups rather than as a hierarchical structure of individual 'reporting' relationships." The process of changing deeply embedded operating assumptions is not easy. First, the existing assumptions need to be surfaced and recognized. Then, people need to be able to experiment with new behaviors rooted in different assumptions, without blame when it doesn't match established norms. Evidence needs to be collected on the results associated with new approaches, but leaders need to advance new approaches well before all the evidence comes in (overcoming the forces of inertia). There will be pivotal test events where the new assumptions are "on the table." It is an iterative process.

For all organizations, many with roots back to the rise of the industrial revolution, the changes in operating assumptions have been essential to keep pace with the first two digital revolutions. The third digital revolution will further challenge the operating assumptions of these organizations. Importantly, many of the operating assumptions that will be on the table are related to a more distributed use of knowledge and skills; that is, increased agency, throughout the enterprises. A more distributed operations model allows for more agile, resilient, and ongoing adaptation, whereas the traditional model is a rate limiter. These organizations will undoubtedly experience a series of pivotal events that will add up to a transformation, just like the Ford Motor Company, but they will not have a window of three decades to make the change.

The challenge doesn't just apply to traditional, industrially structured organizations. As Neil notes, many founding organizations of the first two digital revolutions such as DEC, Wang, Prime, and Data General no longer exist. Just because they were producing digital products did not mean that they had operating assumptions sufficient to match the pace of change in a digital era. In the case of DEC, for example, the rate limiter was partly a failure to address the shift from mainframe computing to personal computing. A closer look reveals additional embedded operating assumptions that were the key to early success and a rate limiter later on. These included a near reverence for the founder, Ken Olsen. Such admiration was great when he was correct in his understanding of the technology frontier and fatal when he was wrong. Additionally, some operating assumptions on the front lines stifled dissent and limited a bottom-up capability to adapt.

Digital organizations that have survived the second digital revolution—such as Microsoft, Google, and Intel—enter the third digital revolution already attuned to exponential rates of change. Each has a distinct culture, with unique combinations of operating assumptions, but all operate with an assumption that the next five years will not necessarily look like the past five years. Intel is a particularly interesting case. Moore's Law is embedded in the DNA of this corporation—its business plan is centered on a doubling of performance every eighteen to twenty-four months. Intel's copy-exactly approach to changes in its operations balances innovation and standardization. Innovations in any one facility are studied and, if they are viable, all facilities are expected to copy the innovations exactly to form a new base on which new innovations can emerge. This operating assumption is a rate accelerator—enabling continuous improvement at the level of the enterprise rather than at an individual-facility level.

These surviving organizations from the second digital revolution are already acting a bit like machines that make machines. That is, the organizations are making new organizations through the continual process of mergers, acquisitions, joint ventures, public-private partnerships, and a wide variety of flexible team-based operating structures. Their operating assumptions not only feature more horizontal forums and mechanisms for planning, decisions, and action, but also function to some degree with the same modular assembly and disassembly that Neil outlines in the underlying digital principles. Still, the analogy is not complete—these organizations will resist complete disassembly. Furthermore, digital fabrication will challenge some operating assumptions for organizations oriented around bits. For example, they will have to consider how their computing technologies interact with active supply chains for consumable materials and intensified social norms around open sharing and barter exchange. Intel is still recovering from its slow entry into the mobile computing market, but the changes associated with the third digital revolution may not have as much room for error.

To examine rate limiters and accelerators at the organizational level, we need to observe rates of change in technology and identify whether an organization's operating assumptions align with the technology. Alignment at the organizational level is not as complex as it is at the institutional level (where many more stakeholders and interests can be at play). On the other hand, organizations that don't adapt can disappear faster than the time it takes to realize that organizational change is not happening.

Individuals

Although some individuals such as technology pioneers and early adopters embrace and keep pace with technological change, not all individuals have this opportunity or even desire. In *Future Shock*, Toffler says that "too much change in too short a period of time" leads to feelings of "helplessness, despair, depression, uncertainty, insecurity, anxiety and burnout." Many people feel this way today, a predictable response to accelerating change.

In the 1990s, a model for managing change with regard to individuals was developed for use in executive education. The *transition curve* depicted in the figure below illustrates how individuals respond to change. There are other similar models in use, most of which have roots in Elisabeth Kübler-Ross's model for dealing with death and dying. In the figure, the vertical axis denotes relative levels of self-perceived competence. The horizontal axis is time. As the curve illustrates, after the initial shock of encountering something new, a person's self-perceived competence increases with his or her denial of key aspects of the change. The process of increasing awareness and acceptance involves acknowledging what you don't know—letting go of old assumptions, perceptions, and other considerations. Only then are experimentation, understanding, and integration possible.

What complicates the transition curve in the context of the third digital revolution is that it has to match the speed of exponential change. This requirement for speed increases the likelihood of denial after the shock and makes the journey to awareness, acceptance, experimentation, and understanding more complicated. William Bridges's 1991 change model in *Managing Transitions* is consistent with the transition curve depicted in

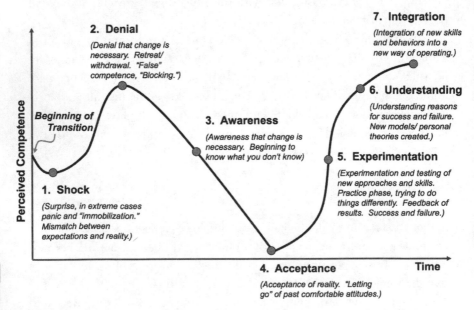

The transition curve. *Ford Motor Company Executive Education, from* Inside the Ford-UAW Transformation

the figure above—pointing out that people have to let go of the old and navigate the "neutral zone" (awareness and acceptance) before they are ready to embrace the new. For individuals facing accelerating rates of change, it will mean managing these transitions more and more frequently.

But for individuals, the implications of accelerating change with the third digital revolution go beyond just a new mind-set. Some very tangible adjustments also have their own rates of change. For example, the time it takes to achieve basic literacy in digital fabrication is approximately six to eighteen months—a relatively fast learning curve. However, deeper mastery of the underlying principles of material science, design thinking, electronic circuit design, and other relevant domains is measured in years, sometimes decades. Advances in the technology will undoubtedly shorten the needed learning curves for basic functionality, just as using a computer or mobile device today doesn't require computer-programming skills to take advantage of the vast array of applications.

The increase in the number of individuals building the needed capability to effectively leverage the technology is depicted in the graph of labs and grads. The upper, solid line plots the same data Neil used to plot the growth of fab labs; the lower, dashed line plots the growth in graduates from the Fab Academy. The vertical scale is both the number of fab labs and the number of people it has trained with the skills to function effectively in a fab lab. The graduates of the Fab Academy often become critical mentors for existing or new fab labs. Of course, there are more ways to

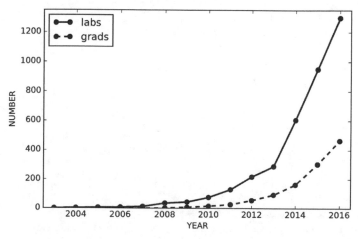

Growth of fab labs and growth in graduates from the Fab Academy. *Neil Gershenfeld and Fab Academy*

learn about digital fabrication than just the Fab Academy, but the figure illustrates both the opportunity and the challenges in cultivating fab literacies and fab mentors. The good news is that there is a growth trajectory of Fab Academy graduates. But not every graduate has the time or desire to mentor the growing number of people who need mentorship throughout the ecosystem. More importantly, the rate of change is slower than the growth of fab labs.

To further illustrate the challenge of the relative rates of change, the following figure extends the two curves above to represent possible future scenarios building on the current data. The top, solid line is a possible growth curve for fab labs at a rate that would put five labs in every city of more than a hundred thousand people (of which there are just over four thousand in the world) and an approximately equivalent number in rural locations. Note that in this scenario, the growth of fab labs begins to level off at the end—this change is the sigmoid that Neil mentioned in Chapter 3 (when it levels off another change comes along in a steeper sigmoid curve). The lower, dashed dotted line of Fab Academy graduates is a conservative estimate on mentoring capability—the training in the Fab Academy. Still, this graph illustrates the risk that the growth of human capability will be a constraint or rate limiter on realizing the potential growth in fab labs. The point here is not to make a specific estimate of growth on either dimension, but instead to visualize the dramatic increase in the rates of change needed in social systems to keep pace with the technology.

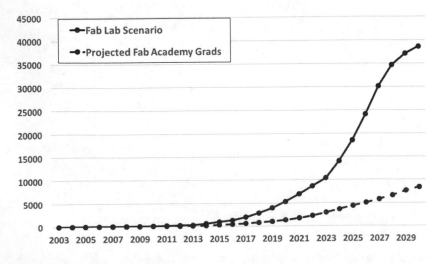

Scenarios of possible growth of fab labs and Fab Academy graduates. *Joel Cutcher-Gershenfeld*

Head, Heart, and Hands

The profiles in this book have primarily been on fab pioneers, individuals who have already bought into the power and promise of digital fabrication. This is still a relatively small group of people. If fab labs are to reach the wider population, it will be essential to engage those unaware or even resistant to the idea of a third digital revolution. As Alvin Toffler pointed out on the cusp of the first two digital revolutions, "future shock is the shattering stress and disorientation that we induce in individuals by subjecting them to too much change in too short a time." The signs of this stress and disorientation are certainly visible throughout society.

For all its potential benefits, the third digital revolution will not be a welcome process for many individuals. Indeed, much of the world is drawing inward, with a combination of isolationist views, distrust of technology, fear of job loss, and anger with growing inequality. Embedded in the concept of digital fabrication is a unique combination of local self-sufficiency and global interdependence that represents an alternative path forward—but that is not an easy concept to communicate in a world of tweets, sound bites, and information echo chambers.

For challenges like fab inclusion to become a global priority, the broader population can't just be asked to embrace more technological change—they need to become stakeholders in helping to shape how the change can positively impact themselves, their families and communities. Social change will succeed better when the beneficiaries become key drivers of the desired change. To address these challenges, we recommend approaches that simultaneously appeal to the head (logic-centric), the heart (emotion-centric), and the hands (practice-centric).

We have all experienced how hard it is to change present behavior to accomplish long-term goals. Whether it's saving money for future financial health, diet and exercise for future physical health, or taking proactive steps to mitigate climate change for the future health of the planet, there is a wealth of evidence that this type of behavior change is extremely difficult. The challenge of engaging people in a third digital revolution is even more of a hurdle because a great many people are still adjusting to the first two digital revolutions.

Neil builds his case for the importance of engaging with the third digital revolution mostly on evidence, logic, and reason. This head-centric approach is also the preferred model for many government agencies and philanthropies that design their theories of change around logic models and input-output analyses. It is absolutely necessary to have a strong, evidence-based analysis of why accelerating digital fabrication technologies can, for example, lead to meaningful education or employment benefits. But this logic centric approach is typically not sufficient if you want to engage a broader population not inclined to immerse themselves in the details.

For many people, it is also necessary to appeal to the heart. For as long as humans have had the ability to communicate, storytelling has been an essential tool for helping one generation guide the next generation through the challenges and complexities of life. Before the invention of writing, these stories were passed down in the form of oral stories, rituals, and plays. The stories that have stood the test of time, those that have transcended cultures and context, are those that explore universal human themes in a highly engaging and relevant format. There is a saying whose origins are not clear (although it is often attributed to the Talmud) that "words which emanate from the heart, enter the heart." Similarly, stories that emanate from the heart enter the heart. Embedding the power and promise of digital fabrication into deeply human narratives, evocatively told, can fire the imagination and serve as a gateway to inspire deeper engagement.

In addition to appealing to the head and the heart, we must also offer hands-on interaction with the evolving technologies. It is one thing to read about a fab lab; it is another thing to get the thrill of harnessing these powerful machines to accomplish personally meaningful goals and then share these accomplishments with friends, family, co-workers, and the broader fab community.

Many of the pioneers of the first two digital revolutions point to their early inspiration as a combination of compelling media (e.g., *Star Trek* television series, Isaac Asimov novels), a few mentors who made the science come to life (e.g., a great science teacher or an inspired parent), and, crucially, hands-on tinkering (e.g., home-brew computer clubs, computer kits). The alchemy across all three of these engagement models is powerful, especially if they reinforce each other.

One important note: the early home-brew computer clubs and computer-building kits in the 1980s were very much the domain of nerdy white males. That stereotype persists today. There is currently much more diversity in the fab ecosystem (as is evident in many of the case studies highlighted in this book). This inclusiveness needs to be highlighted to continue to attract diverse new stakeholders.

There is actually a long history of connecting head, heart, and hands. Aristotle connected the virtuous heart with the head, the mind of a philosopher. Religious organizations over the years have connected the concepts through the integration of thought, passion, and action. Psychologists use a similar framework when they speak of cognitive, affective, and behavioral change. These integrated elements can also be found in numerous modern organizations, such as 4H Clubs (head, hands, heart, and health) and even, almost, MIT. The official MIT seal, adopted in 1864, features the Latin phrase *Mens et Manus,* "the head and the hands." What is missing is the heart.

This lack of heart in the MIT motto has not gone unnoticed by students, faculty, and alumni. In September 2015, following a tragic series of student suicides, MIT chancellor Cynthia Barnhart and medical director William Kettyle announced the MindHandHeart Initiative to improve the coordination of MIT's existing mental health and counseling offerings. In May 2016, at the hundredth-anniversary celebrations of MIT's presence in Cambridge, MIT alumnus and *Car Talk* host Ray Magliozzi, in partnership with a student-built robot, publicly came to the conclusion that it has only been by adding *cor,* or "heart," to mind and hands that MIT has developed into the thriving institution it is today. For individuals in the third digital revolution, simultaneously engaging the head, heart, and hands can be a rate accelerator; not doing so will be a rate limiter.

ECOSYSTEMS

We have examined how rates of change for individuals, organizations, and institutions can be rate limiters. Although it is important to understand each level separately, ultimately any meaningful change involves understanding how all three interact with each other. If digital fabrication technologies are to accelerate at exponential rates of change and help catalyze a more self-sufficient, interconnected, and sustainable society, there will need to be ecosystem-level transformation—regionally and globally.

In biology, an ecosystem consists of interacting organisms and their environment. Co-evolution in biology is a product of the interactions. When it comes to technology and society, the ecosystems include the interactions between individuals, organizations, and institutions with technology and the natural environment. Co-evolution happens, sometimes planned and sometimes unplanned. To anticipate the challenges and opportunities for making societal-level ecosystem change, it is helpful to look back at a few historical attempts at alternative ecosystems.

Relatively early in the industrial revolution, Robert Owen's story illustrates an ecosystems approach that might have been an alternative economic model in society. Born in Wales in 1771, Owen rose to be the manager of a textile firm in Manchester, England, by the age of twenty-one. By 1799, as co-owner of the Scottish New Lanark mill, Owen rejected the way mills were then being run and constructed what we would now call an ecosystem approach to social change. The effort included infant child care in the community, support of schooling for the workers' children, safe working conditions, fair pay, worker education, worker feedback on the quality of goods produced, and cooperative retail outlets so that workers would not have to buy "shoddy goods." Imagine if this these changes were taken up by other entrepreneurs and influencers throughout society—it would have redefined the industrial revolution in many beneficial ways.

In fact, Owen's model ran headlong into resistance from what was by then the established industrial order. Other co-owners of the mill protested that not enough of the profits were coming their way. There was apparently already an expectation that the business should maximize only profits. For Owen's mill, the matter was resolved with the help of social reformer Jeremy Bentham and others, who bought out the protesting co-owners. This extended the life of the experiment, but was not an easily replicated practice in other settings. The New Lanark mill did become a magnet for what we would today call benchmarking visits by social reformers, church leaders, politicians, and others, which increased the potential for a broader impact.

As it turns out, the aspects of the model associated with mitigating harm did impact society in some significant ways, but the aspects associated with creating value did not. Owen became a driving force in the passage of the 1819 Cotton Mills and Factories Act, which focused on child labor and working hours. He led the movement for the eight-hour working day, with the motto "Eight hours labour, Eight hours recreation, Eight hours rest." In this regard, the protective legislation could be seen as a rate accelerator for mitigating harm in a era of technological change.

What could not be as easily legislated and enforced, however, were the aspects of the model associated with creating value, such as worker consultation, cooperative retail outlets, promoting quality, and other social innovations. Although Owen documented the lessons and sought to promote them in his home country and in the United States, the combination of benchmarking and his book were not enough to get a critical mass of people adopting, adapting, and extending this alternative model.

If we fast-forward to the 1950s, we find another alternative ecosystem model that was able to regionally scale for both creating value and mitigating harm. The Mondragon cooperative in the Basque region of Spain was founded in 1956 by a Catholic priest. The cooperative has grown to employ over seventy-five thousand individuals in hundreds of small and medium-sized firms. The model is different than the traditional industrial model in some key ways. If a firm fails, for example, income support and retraining is provided, along with investment opportunities for people who want to launch new firms. Cluster del Conocimiento is a collection of working groups that serve as a knowledge engine for Mondragon, looking ahead to new business opportunities and changing markets. Key social services are available to all. Within this context, individual, organizational, and institutional changes have come together to enable internal growth and long-term sustainability.

Even though it is a product of the relatively closed Basque culture, Mondragon has welcomed tens of thousands of benchmarking visits and been the subject of dozens of books with the aim of extending the model to other settings. Today, it is working to be effective in a digital world. It is perhaps no surprise that it is a receptive context for digital fabrication; it is dedicating research and development resources to develop new materials for 3D printing (releasing a new nylon filament for additive manufacturing in 2015). Still, Mondragon remains a relatively closed system that interacts with the external markets primarily to sell what it produces and has not made a deep societal impact beyond its region.

In discussing Mondragon, Tomas Diez, who is leading the Fab City initiative in Barcelona, points out that there are important insights to be garnered from the Mondragon model but also differences when it comes to the Fab City ecosystem. He emphasizes a clear distinction between the self-contained, relatively closed model of the Mondragon cooperative and the lateral, collaborative approach that is needed for the third digital revolution: "Mondragon is a different model—we don't aim to be brokers for everything. We have a role to play, but we are more enabling. We are supporting a growing ecosystem of fab cities, but we do not need to be in the center." In these comments, Diez is articulating the approach of a distributed ecosystem designed to propagate on a global scale.

A more recent model for ecosystem change, with a focus on transforming education, is the Remake Learning initiative, covering the Pittsburgh region as well as Western Pennsylvania and Northern West Virginia. Remake Learning is a collaboration of more than two hundred organizations expanding opportunities and enhancing learning outcomes for young people in the region. It aims to inspire a generation of lifelong learners, preparing them to thrive in the twenty-first century. The model includes an estimated 150 maker spaces and fab labs.

The story began twenty years ago, when the Manchester Craftsmen's Guild led revitalization efforts in a long-neglected Pittsburgh neighborhood by connecting at-risk youth to apprenticeship training, arts education, and other forms of out-of-school learning. A decade later, in 2007, Pittsburgh's Grable Foundation launched Kids+Creativity in response to a seemingly simple concern that was surfacing from teachers, librarians, museum educators, and youth workers: "I'm not connecting with kids the way I used to." At first, this concern seemed no different from the refrain of every generation with respect to youth. But further investigation revealed that the pursuit of knowledge was emerging as different in important ways in the digital age.

Kids+Creativity began with a unique practice—ten pancake breakfasts bringing together leaders from schools, museums, libraries, early-learning centers, and out-of-school programs. Unsolicited, participants in the breakfasts each said they knew two or three people who also needed to be part of the conversation. Following these breakfasts was a larger gathering of over one hundred people. As described by Grable Foundation executive director, Gregg Behr, the meeting was "like The Gong Show—everyone had two minutes to make their comment or pitch, and then there would be a ringing of the gong indicating time was up. What everyone said was very powerful." Additional funding and support followed from many sources, including the Claude Worthington Benedum Foundation, the Buhl Foundation, the McCune Foundation, the Pittsburgh Foundation, the Sprout Fund, and the Allegheny Intermediate Unit (an educational service agency). Pilot experiments were launched involving project-based learning, maker and fab-based learning, game-based-learning, and other innovative approaches to teaching and learning—often spanning what were previously separate domains—in schools, libraries, museums, out-of-school programs in a mix of formal and informal learning contexts.

Rebranded as Remake Learning in 2011, the initiative has drawn national attention and additional support from the John D. and Catherine T. MacArthur Foundation. The umbrella term remake learning was deliberately selected to be descriptive, without choosing from among the many contending terms and pedagogies emerging among educational innovators and ed-tech evangelists. In 2015, Chevron joined with the Carnegie Science Center and the Fab Foundation to launch a comprehensive fab lab in the region (as part of a $10 million commitment by Chevron to the Fab

Foundation to support fab labs in areas where Chevron has operations). This fab lab connects with the estimated 150 maker spaces in the Greater Pittsburgh area, linked in various ways to the Remake Learning network, including the TechShop learning center, HackPittsburgh, the Western Pennsylvania School for the Deaf's Makerspace, Maker Faire Pittsburgh, and many others.

Interestingly, the initiative was seven years old, in 2014, before it formed a top-level leadership council (consisting of thirty-six regional senior executive leaders). This reflects the early emergent nature of the initiative and also points out the need for governance as an enabling practice in an ecosystem. There is now a full-time professional staff and an interconnected network of two thousand educators and community activists, nearly 40 percent of whom have joined within the past two years. Pittsburgh is one of thirty digital education clusters in the United States and, arguably, the most advanced. The level of outside interest has led the Remake Learning network to develop a playbook that summarizes lessons learned in five domains:

- Learning environments
- Innovation research and development
- Learning scholarship and advocacy
- Commercial and entrepreneurial engagement
- Strategic stewardship

The playbook represents an explicit effort to help make transformational change beyond one region. While primarily analog approaches to scaling change—the writing and publishing of books (and playbooks), the organization of site visits and benchmarking, pursuing changes of legislation—are all important activities, they are ultimately limited in their ability to accelerate change at a pace to effectively co-evolve with the accelerating technologies. In order for this to happen, there will need to be a continual honing of the digital platforms, tools and practices so that key aspects of the model can effectively propagate. For example, while there are dozens of maker spaces and fab labs throughout the region, there is still limited ability to share and collaborate on projects across the network. Dr. Todd Keruskin first became involved in Remake Learning as a high school principal and now serves as assistant superintendent of the Elizabeth Forward School District. He comments:

For digital fabrication, the key is creative projects. The Fab Foundation said it would develop a platform to share projects and I am told it will soon be announced. This is good. We need effective design challenges—particularly project ideas to help the community—but we have struggled to develop them internally.

Here there is a need for new practices—challenging and beneficial projects—with the associated tools that can reside on a platform where they can be adopted, adapted, and extended. Since most of these practices will not be unique to the Pittsburgh area, there will be the potential to propagate beyond the region.

In the case of Remake Learning, digital fabrication is part of a new teaching and learning ecosystem. Still emerging is the regional capacity for digital fabrication platforms, tools and practices to create more self-sustainable communities with new models for how we live, learn, work and play. To help accelerate this process it is is helpful to look at how digital technologies have enabled a single individual or a small group of people to create truly transformative ecosystems with global impact.

PROPAGATE VERSUS SCALE

The recent history of the first two digital revolutions offers insights into designing and propagating transformative ecosystems. It is now possible for a single individual or small group of passionate visionaries to engage and empower millions of people to contribute to the design and propagation of globally transformative ecosystems. These digital ecosystems leverage passionate networks of contributors, rest on powerful platforms and tools enabling distributed agency, and encourage practices that are capable of propagation at exponential rates, not just increases on a linear scale—offering unprecedented rates of change.

In Chapter 2, we identified the cultivation of an enabling fab ecosystem as a key threshold challenge to creating a more self-sufficient, globally connected and sustainable world. Here we go deeper into this concept, including how the properties of these ecosystems are in a constantly emergent state, which is essential for exponential propagation. Emergent ecosystems share some similarities with how traditional businesses scale, but there are also key differences. Businesses can scale through a combination of planned implementation, multi-territory research and design, supply change management, customer support, international management, along with mergers and acquisitions. Scaling in this traditional way is often very capital intensive and takes experienced, sophisticated management. This is more of a linear process with periodic accelerations.

Note that there are two key words here—propagate and scale—which are used in very different ways in Neil's world and ours. We tripped over this in writing this section and it is important to spell out the differences since each word is used in almost opposite ways in our different worlds. For Neil, physical propagation, such as with waves, happens with a defined velocity from a given source that is in contrast with the doubling that is at the exponential heart of scaling Moore's Law. In our world, both for-profit and non-profit organizations attempt to scale their operations with a variety of mechanisms that mostly involve top-down planning and implementation. This contrasts with what is almost an organic process of propagation that we explore in this section.

Emergent ecosystems, in the way we are using the terms, don't scale; they propagate. The platforms, tools and practices are not the same as an overarching hierarchy. Instead, structurally, they involve a form of lateral alignment—connecting independent but interdependent stakeholders. For social systems to match the exponential growth possible with technical systems, they need to propagate (often in addition to traditional scaling).

A few years ago, Alan attended a forum on adaptability at the Biosphere 2 near Tucson, Arizona, where this topic was explored. The convening brought together researchers across a variety of science, social science, and humanities disciplines to explore strategies for adaptation in complex, rapidly changing ecosystems. One of the group exercises involved looking at transformative ecosystems from the first two digital revolutions and deconstructing the elements that helped them propagate and remain resilient in a world of constant and accelerating change. The group included a diverse mix of biologists, economists, philosophers, and entrepreneurs. After the conference, a small subset including John Abele, Boston Scientific co-founder and former chairman of FIRST Robots; Ken Parker, CEO of NextThought; and Alan began documenting some of these insights.

The group started with the assumption that humans have emerged as a dominant species in large part because of our ability to collaborate in large numbers. Digital technologies have accelerated this ability enabling globally transformative ecosystems to emerge in a remarkably short amount of time—largely because of shared and distributed agency across aligned communities—united by core platforms, tools and practices.

Examples of such transformative ecosystems range from decentralized technology movements (e.g., Linux) to for-profit entertainment (e.g., Minecraft) to nonprofit web resources (e.g., Wikipedia).

These three examples—Wikipedia, Linux, and Minecraft—have had a considerable impact on the world. Wikipedia is the world's sixth-most popular website with a global ecosystem of contributors. It is a primary source of information for hundreds of millions of people, with close to five hundred million monthly readers and eighteen billion page views. A recent report by the Linux Foundation estimated that Linux software has generated $5 billion in economic value through 115 million lines of code contributed by a global ecosystem of coders. And Minecraft has emerged from a small, independent game created by a single Swedish game designer to a global ecosystem with more than a hundred million passionate and deeply engaged youth accelerated through a global ecosystem of 'modders' who adapt and extend the code to introduce new experiences. Each has grown through propagation rather than the more traditional forms of scaling.

All three examples, along with others across a variety of domains, also share a common set of attributes. The ecosystems were founded by a charismatic and passionate leader or small group of leaders who defined the community ethos, culture, and high-level rules of engagement (either overtly or through their behavior and practices) and these founders provided the enabling platforms, tools and practices that empowered a growing ecosystem. In some cases, these founders set out to create transformative ecosystems. In other cases, they did not aim this way and were hence unprepared to be thrust into such a leadership position. We'll call these passionate leaders the top-down visionaries. In the fab ecosystem, Neil certainly fits into this category.

Each of these ecosystems have grown in large part through the passion of an empowered and empowering middle tier of individuals and organizations, all of whom deeply identify with the vision and emerging community—and effectively leverage, adapt, and extend the platforms, tools and practices. The empowered and empowering middle devote considerable time and energy (often with no compensation or formal role) to adapting and extending the platforms and practices to meet local needs or create new capabilities. Their dedication, in turn, brings diverse new participants into the ecosystem raising their stature in the community. Influence comes mostly through action and tangible results benefitting the community, not by top-down decree. E-Line Media and Teacher Gaming's work helping to bring the Minecraft mod (MinecraftEdu) to thousands of schools is an example. We'll call this tier the empowered and empowering middle. In the fab ecosystem, these are the passionate founders of fab labs, the coders and designers who contribute to the open-source fab software and hardware development, and those who pioneer new initiatives like the Fab Academy, the Fab Foundation, and the Fab City movement.

Last, but certainly not least, all of these ecosystems attract and benefit numerous participants who engage to varying degrees. We'll call these stakeholders the bottom-up participants. In each of these emergent ecosystems, there is agency for stakeholders at all tiers, with appropriate levels of responsibility and ability at each tier—all aligned by shared platforms, tools and practices as well as, often, a shared culture and vision. In the third digital revolution these are the growing number of engaged participants who interact with the global fab ecosystem.

The founding visionaries of these products and services have been celebrated throughout the media, as have the millions of engaged bottom-up participants in these ecosystems (sometimes referred to as the "wisdom of the crowd"). The idea that these ecosystems are either top-down or bottom-up driven, however, is a false dichotomy. It is missing the importance of the critical middle tier that often represents the secret sauce for how these ecosystems propagate in an emergent, decentralized way. Much like thriving ecosystems in nature, this middle tier creates diversity, redundancy, and adaptation, enabling the ecosystems to continually adjust to changing environments and become a force multiplier in how they propagate. Examples of this empowered and empowering middle tier include those who modify Minecraft, the editors of and contributors to Wikipedia, the Linux and Mozilla developers—to name a few.

Interestingly, not all of the founders of these emergent ecosystems intended or even desired to create a movement or change the world. The best example is Markus "Notch" Persson, the creator of Minecraft, who just wanted to make a fun game. When he was thrust into the role of a virtual messiah for tens of millions of tweens, the situation became simply too much for him. Out of the blue, he sold the game for $2.5 billion to Microsoft, the last thing anyone following his active dialogue with the community thought he would ever do. In an open letter to his fans, he explained his reasons: "I've become a symbol. I don't want to be a symbol, responsible for something huge that I don't understand, that I don't want to work on, that keeps coming back to me. I'm not an entrepreneur. I'm not a CEO. I'm a nerdy computer programmer who likes to have opinions on Twitter. . . . If I ever accidentally make something that seems to gain traction, I'll probably abandon it immediately. . . . It's not about the money. It's about my sanity."

Most of the emergent ecosystems that have reached tens of millions of people have centered on software. A key question for platforms that cross from bits to atoms is the impact that the atoms have on this ability to rapidly propagate. This challenge was underscored by Mitch Resnick, who directs the Lifelong Kindergarten Group at the MIT Media Lab and is the creator of the globally popular youth computer-programming platform Scratch, launched in 2007. Resnick came up with the idea for Scratch while developing creative learning experiences for the Intel Computer Clubhouses. The website is now getting one hundred million unique visitors per month, and more than twenty-one million Scratch projects have been developed. Resnick points out, "When Intel announced its support for expanding the Computer Clubhouse network, I warned them that community centers couldn't expand according to Moore's Law. That was right. But the Clubhouses gave birth to Scratch, which has been growing according to Moore's Law."

That said, there are physical-world ecosystems that also fit the emergent ecosystems model, though (as with the fab labs) the accelerating growth has been slower than the leading digital ecosystems. A good example is FIRST robotics, an international high school robotics competition. FIRST was co-founded by inventor Dean Kamen in 1989 to "create a world where science and technology are celebrated . . . where young people become science and technology heroes." FIRST has reached more than three hundred thousand middle and high school youth and has over ninety thousand volunteers. Kamen is a visionary founder, and the distributed network of deeply committed volunteers is truly an empowered and empowering middle tier and force multiplier in the ecosystem. One of the most impressive things about FIRST robotics is how the culture of gracious professional teamwork and co-opetition has permeated the entire distributed ecosystem. The idea for gracious professionalism came from Woody Flowers, an emeritus professor of engineering from MIT. Flowers helped build the FIRST ecosystem and now serves as national adviser to the organization. Anyone who has been to a final national FIRST competition (which often fills entire stadiums) knows that the community has managed to infuse a competitive environment with a culture of cooperation and respect throughout the ecosystem.

A key question, then, is whether the fab ecosystem, with its mix of digital and physical elements, can propagate into a globally transformative ecosystem. A second question is whether such an emergent ecosystem can truly lay the foundation for a more self-sufficient, interconnected and sustainable world. Through a blend of thoughtfully architected fab platforms, tools and practices it is now possible to empower a global, distributed network of passionate individuals with aligned interests and shared objectives. Inspired path creators can now not only design innovative new social and economic models, they also have the ability to propagate these models through globally transformative ecosystems. But, before the social systems can truly evolve with accelerating technology, it is first necessary to have a deeper understanding of the projected technology roadmap, the subject of Chapter 5.

CHAPTER 5

The Roadmap

Envisioning the future impact of digital fabrication has been a science fiction staple, even if it wasn't always called that. In the *Star Trek* TV series, the replicator was a handy device that could produce any required story element, which more often than not meant Captain Picard requesting tea, Earl Grey, hot. Somewhat more ominously in the *Terminator* movies, the T-1000 was the most fearsome opponent, a shape-shifting liquid-metal man that could take on the appearance of anyone, and its droplets could reassemble when they were inevitably blown apart. These sci-fi inventions portray the conclusion of the third digital revolution, in universal replicators and programmable materials (hopefully minus the Skynet part).

To realize this technological vision, in 2013 I ran a workshop with the White House to examine the research roadmap. This event was organized in response to interest across federal agencies in formulating their plans for 3D printing. I felt that these agencies were missing both the range of technologies for digital fabrication in the short term and the emerging science of fabrication in the long term. The workshop identified four stages:

- Using computers to control machines that make things (as are found in a fab lab today)
- Rapid prototyping of rapid-prototyping machines (fab labs making fab labs)
- Coding the construction of digital materials (merging all the machines in a fab lab into a single process)
- Programming materials (merging machines and materials)

These stages have a natural correspondence with the exponential projection of Lass' Law:

- The first phase, the one we're currently in, is coming to an end as the number of fab labs becomes comparable to the number of cities on the planet—a few thousand of each, for an average of one fab lab per city.
- The second phase is an increase by a factor of a thousand, going from the thousand fab labs now to the equivalent of a million. That's on the order of the number of local governments on the planet, making the capabilities of a fab lab widely available to individuals as well as organizations.
- The third phase is another increase by a factor of a thousand, from a million to the equivalent of a billion fab labs, which is the order of magnitude of the number of people on the planet. In the same way that computers and cell phones have become ubiquitous, the third phase is when access to the capabilities of a fab lab becomes not just personal but universal.
- The fourth phase goes with another factor of a thousand, from a billion to the equivalent of a trillion fab labs. Along with counting people, a billion is also the number of computers connected to the Internet before the original addresses (IPv4) ran out. Taking one thousand as an estimate of the average number of things a person possesses, a trillion is thus on the order of the total number of things people might interact with or the number of things computers are connected to. This is when digital fabrication makes not just almost anything, but also almost everything.

Lass' Law began by counting fab labs, which could be defined as containing the core set of technical capabilities to make (almost) anything: designing and scanning in 2D and 3D, fabricating with additive and subtractive processes, creating and interfacing circuits, embedding and programming computing. As this scaling progresses, we count no longer fab labs as they exist today, but rather access to the equivalent capability to fabricate physical forms and program their functions. What's changing is the inputs rather than the outputs: a fab lab today rests on a global supply chain to source things like integrated circuits, precision tooling, and polymer resins. Over time, the outputs will be produced within the descendants of today's fab labs, not by requiring massive upstream capital investments as is currently done, but by assembling ever-more-fundamental building blocks.

There are two ways to project how long this transition will take. The more conservative approach is to assume that Lass' Law continues to hold, with a doubling time of a year and a half. A factor of a thousand is then about ten doublings, or fifteen years. From where we are now, that's forty-five years, the same run Moore's Law has had. The more ambitious projection recognizes that the technologies for all four stages can be seen in the lab today, and it's just a question of how quickly they can get out the door. We can't answer which projection will hold, but we can provide a tour of each of these stages.

COMMUNITY FABRICATION: 1 TO 1,000

The first phase on the digital fabrication roadmap starts with the tools found in a fab lab today. This stage belongs in a chapter on the future because, as the author William Gibson has observed, the future is here today; it's just not very evenly distributed. Because many more people haven't yet been in a fab lab than have, the first step in understanding the roadmap is seeing where it starts.

Cutting

Cutting tools move in two dimensions to cut out parts from flat sheets. Although that might sound to be of limited utility, cutting can be much faster than the other processes. Cutting tools can work with materials that the other tools can't handle, and kits of these parts can be rapidly put together to make 3D objects.

By far, the most popular tool in a fab lab is a laser cutter. Its only input (other than the stock to cut) is electricity, and its only output (other than the cut parts) is exhaust gases from the cutting process. In minutes, a laser cutter can plot out complex shapes with features as fine as the laser beam, a few thousandths of an inch. A thousand-dollar laser cutter can cut thin sheets of wood, cardboard, and plastic. A ten-thousand-dollar one can cut larger, thicker sheets, and a hundred-thousand-dollar cutter has a much larger bed and can cut metal.

The most versatile and least appreciated cutting tool moves a powered knife, commonly called a vinyl cutter, because its most common use is to

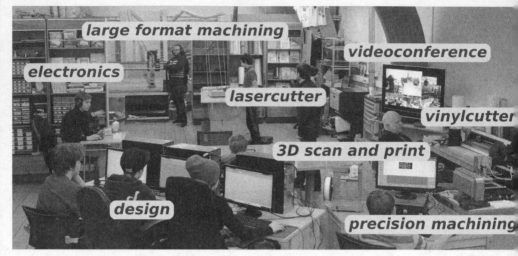

large format machining

videoconference

electronics

lasercutter

vinylcutter

3D scan and print

design

precision machining

Fab lab tools. *Frosti Gíslason*

cut out vinyl lettering for signs or stickers. But these tools can also cut out masks for screen printing, score *origami* folds, add *kirigami* cuts, and plot flexible circuits and antennas in conductors. The largest ones can handle large cardboard sheets to cut and fold display cases and furniture.

The most powerful of these tools is a water-jet cutter. It shoots a fine, supersonic jet of water that carries particles of a garnet abrasive. Because the particles are traveling so fast, they blast through anything they encounter—thick steel, glass, ceramic, stone. With such versatility, water-jet cutters are taking over the work of other more specialized tools. The downside is that they need to be supplied with a steady stream of the abrasive, which produces an equally steady stream of waste to be disposed.

The last group of cutting tools uses a wire. One kind is heated and can melt through soft materials like foams. In electrical discharge machining, a thin wire is energized with a current that makes tiny sparks that can erode deep cuts in the hardest metals. For that reason, electrical discharge machining is the tool of choice to make the precision parts required for machines to make machines.

Milling

Milling machines move a rotating cutting tool in three rather than two dimensions, and more advanced ones can tilt the tool or workpiece as well to simultaneously control four or five axes. These machines range in size from a machine that fits on a desktop to a machine that's the size of a house. And their corresponding cutting torque ranges from that of a small hand drill to that of a large engine.

Milling machines are noisy and messy. The cutting process and the motors that drive it are loud. The material that's removed flies off as chips that must be captured and disposed of. Depending on the material, it can be necessary to spray a lubricant over everything. And they require an assortment of tooling that needs to be regularly replaced.

But in return for all of that, milling machines can make smooth surfaces in hard materials with precise features. That's why they're the tool of choice for making things like automobile and aircraft components, high-end laptop cases, custom furniture, and tools used in other production processes. Smaller, higher-precision milling machines are used for jobs like cutting out the traces in a circuit board and making molds to cast parts.

Printing

3D printers deposit rather than remove material. Additive fabrication has two important differences from subtractive fabrication. Consider making a ball-and-socket joint. A milling machine would have to carve out the ball, then the hemispheres of the socket, which would have to be assembled around the ball. Because the 3D printer can build the part up in layers, it has access to what will become the inside as well as the outside, so that it can make the ball in the socket in one go. And it can deposit material only where the parts are, so there is no scrap wasted, other than possibly some extra material used to build supports.

The first 3D printing was done in 1986 through *stereolithography*, a process that scans a laser beam to selectively solidify a layer of a polymer liquid that's cured by light. Stereolithography is still used to make parts with the finest features and smoothest surfaces, because the resolution is determined by the size of the laser beam. The downside is that the result comes out as a gooey mess of cured and uncured polymer that needs to be cleaned, and the mechanical properties are limited to photopolymers. The second approach, fused deposition molding, became available in 1989. The printer feeds a polymer filament into a heated nozzle, which extrudes a thin bead that solidifies on the growing part. That's not as fine as a laser beam, but the process can now use stronger thermoplastics and is much less messy. Fused deposition molding is most common for entry-level thousand-dollar 3D printers, because the process is so easy to implement. A more recent approach uses a printer head similar to an inkjet printer's, but instead of shooting droplets of colored ink, it deposits droplets of a polymer. Just as an inkjet printer can mix multiple colors, these can mix multiple materials. And the most expensive 3D printers use a much more powerful laser to selectively sinter layers of a fine powder; these printers can now make strong metal parts and tough, high-temperature ceramic ones. The downside to these printers is that they're so complex they can cost a million dollars and require a dedicated room and technician to keep them running.

Although 3D printers are good at creating complex geometries, with nested features that can't be reached subtractively, these tools are slow, typically taking from hours to even days to make something. And the materials in the form that printers use them are much more expensive than the bulk materials. Other approaches can be bigger, faster, cheaper, or stronger. That's why, when all the other digital fabrication tools are available, 3D printers are used only for the jobs they're best suited for. In the CBA shop that I run, that might be a quarter of the time.

Scanning

The inverse of 3D printing is 3D scanning, the creation of a digital model from a physical object. Reasons for scanning include copying objects that don't have digital designs, sculpting physical objects as part of design workflows, creating realistic assets for computer graphics, and conserving valuable artifacts.

The simplest kind of 3D scanner is just a camera. With pictures taken from many directions, photogrammetry algorithms can solve for an object's 3D geometry using just the collection of 2D images. Photogrammetry doesn't need specialized equipment and can be done anywhere, but can be confused by visual artifacts and gives only an approximate estimate of the exact shape. More accurate scanning can be done by laser scanners that follow the position of a laser beam as it's scanned across an object, and by structured light scanners that project an illumination pattern on an object. But these methods still have trouble with visual artifacts, such as a shiny, smooth surface that reflects the illumination away from the detector. Visual artifacts can be fixed in a light stage, which can simultaneously illuminate and collect light from all directions. Light stages determine not just geometry but also all the optical properties required for a computer graphic model, but require much more substantial installations. The most advanced 3D scanners use X-rays to reconstruct the inside as well as the outside of objects, taking many projections from different directions in a computed tomography scanner. These scanners are also the most expensive, with costs ranging over a million dollars. Nice if you can afford it.

Molding

These techniques don't fit within a division between additive and subtractive processes, because they encompass aspects of both. But molding tools are responsible for most of the mass-produced products around you. Digital fabrication is direct-write, meaning that everything it makes can be different. But if you do want more than one of the same thing, digital fabrication can be used to make a tool to make multiple copies of something, saving time and money per part as well as offering enhanced surface finishes and material properties.

Molding processes start by making a mold by numerically controlled (NC) machining or by cutting and folding. They can also be 3D printed, but it's typically faster to remove material, and molds made that way can be smoother and stronger. Materials can then be poured, injected, inflated, drawn, spun, or pressed into the mold. This is done with plastic, metal, foam, food, glass, and concrete. The total time to make a mold is comparable to the time to make a 3D print, but each subsequent use of the mold can take minutes or even seconds, depending on the material. Mold-making used to require large production runs to justify the expense, but with the advent of low-cost NC mills, molds are now feasible for short-run production.

For making larger lighter things, the molds can be filled with fibers that are strong in tension, embedded in a resin matrix that is strong in compression, to produce composite parts. Carbon fiber, fiberglass, and natural fibers are commonly used, along with epoxy, plastics, and natural resins for the matrix. Composites are being rapidly adopted for saving weight, eliminating parts, and improving performance in cars, airplanes, prosthetics, and all sorts of sports equipment.

Computing

All the preceding tools are based on digital fabrication in the original sense of using computers to design things and then to control machines to make them. This last group completes that connection by embedding computation into fabrication.

What has reduced the size and cost of rapid-prototyping tools is the drop in the size and cost of the computers that control their actuators, read their sensors, interpret commands, interface with users, and communicate with networks. Those same capabilities are now appearing in a range of formerly inanimate objects that are becoming part of an Internet of Things, such as intelligent infrastructure in buildings to save energy, and unobtrusive medical monitoring to improve health care.

Hobbyists today build intelligence into their rapid-prototyping projects with small single-board computers with names like the Arduino and the Raspberry Pi. These cost tens of dollars but are composed of even smaller integrated circuits called microcontrollers, which are simple but complete computer systems on a chip. Microcontrollers range in cost from dollars down to tens of cents.

The circuit boards that electronic components are attached to are mass-produced by chemical etching, but this process creates hazardous waste that must be collected and safely disposed of. For short-run production, circuit boards can be made directly with digital fabrication tools: milling machines can carve the traces from copper-clad boards, vinyl cutters can plot them in flexible copper sheets, laser cutters with the right optical properties can ablate them, and certain kinds of 3D printers can extrude conducting material. The electronic components can then be attached by hand with surface-mount rework tools, and for larger numbers and larger boards, they can be attached automatically with a special kind of robotic printer called a pick and place, which takes the components from tape reels and positions them on the board.

With the ability to rapidly make programmable circuits added to rapid physical fabrication, it's possible to produce complete functional systems, as we saw in Chapter 1. In the sections that follow, the change is not what can be made; it's what's required to do it, replacing first the roomful of tools and then the inventories of materials and parts.

PERSONAL FABRICATION: 1,000 TO 1,000,000

The second phase in the roadmap is defined by the ability of a fab lab to make another fab lab. This will be accomplished not by reproducing the machines as they exist today, but by merging their capabilities with modular designs and constructions. Eliminating the need to acquire (and house) many separate machines corresponds to the period when PCs arrived in the history of computing. The PCs were initially used by early adopters and later spread throughout society.

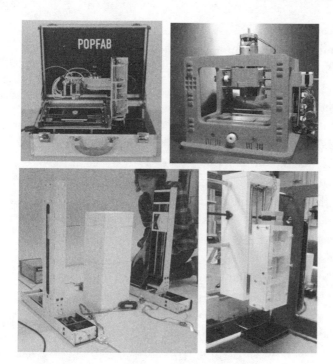

Fab lab machines made with fab lab machines. *Nadya Peek*

The first factor of a thousand has the natural interpretation of roughly one fab lab per city on average. The natural extrapolation for this next phase is a thousand times a thousand fab labs, equipping every community of any size with the means for local production. The sales of 3D printers is already at hundreds of thousands a year and in a few years is projected to reach millions. The obvious conclusion is that millions of all the other tools currently in a fab lab will be sold also, so that the full set is available everywhere.

Such a scenario would present two problems. At today's prices, the cost of that many copies of that many kinds of machines would add up to a hundred-billion-dollar investment, an amount approaching the size of major components of the global economy. And along with finding that much money, it would be necessary to find a corresponding amount of space for all the rooms that those machines would fill.

The straightforward solution to the first problem mirrors the solution to the same problem in the first and second digital revolutions: economies of scale. The bill of materials for rapid-prototyping machines can be as low as hundreds of dollars; they currently cost much more than that because of the need to recoup development costs and sustain the small businesses that make them. A number of these small businesses aim to become big companies, mass-manufacturing these machines for a consumer rather than niche market, much as IBM and Apple brought the manufacturing of personal computers to scale.

But this straightforward solution of economies of scale doesn't consider that this time around, there's an alternative: the technical goal of a fab lab has always been to make another fab lab. If what's being deployed is machines that can make more machines, then it's not a good business plan to make the machines. Instead of mass-producing the machines, you can mass-produce their components and assemble then locally. What I had initially missed was how this idea solves the second problem—that is, the amount of space that these machines occupy—by transforming the concept of what a machine is.

To help drive the transition from buying to making a fab lab, in 2004 I began periodically following my rapid-prototyping class, How to Make (almost) Anything, with one on machine building, How to Make Something That Makes (almost) Anything. Two students, Nadya Peek and Jonathan Ward, made a machine called the MTM Snap in the class (*MTM* for "Machines That Make," and *Snap* because instead of using fasteners to hold the machine together, it is snapped together from parts milled in high-density polyethylene, the dense plastic used to make kitchen cutting boards). The MTM Snap was a precision tabletop milling machine, with specifications similar to a few-thousand-dollar machine in the fab lab inventory, but at about a tenth of the cost. Once we posted plans for it so that the machine could be made in any fab lab, I thought our work was done.

It wasn't. It turned out that it could be made by anyone in a fab lab, as long as you were Nadya or Jonathan. Too much knowledge and skill were required for most people to be able to successfully reproduce it, so there were many failed attempts. Instead, the design was spun off and commercialized as a finished product by the Other Machine Co. A variant of the design became ShopBot's Handibot machine. Another alum from the class, Max Lobovsky, started the company Formlabs to sell high-resolution consumer stereolithography 3D printers. Machine building at a fab lab in the Netherlands led to the Ultimaker family of fused deposition molding 3D printers.

Although each of these machines was faster, better, and cheaper than its predecessors, all the machines do share an implicit assumption: they do what they're designed to do. This observation might sound tautological, but a close analogy with the history of computer software and networks explains both how to simplify their construction and how to eliminate the need for so many different types of machines.

Software was originally written as programs that did something, then a different program would be written to do something else. That proved to be very inefficient, because there was very little reuse of code. It's no longer done that way. Instead, what's called object-oriented software is now written in reusable packages that combine routines that perform a function, the data that they operate on, and the inputs and outputs to use them. These software objects can be composed to do larger tasks, so that, for example, if you're starting a website, one of these could serve the web pages, one could perform electronic commerce transactions, and one could keep track of a user's identity.

Computer networks were originally developed with a task in mind, such as running a factory or banking. These special-purpose networks have been largely replaced by the general-purpose Internet, which has as one of its core architectural principles what's come to be known as the *end-to-end argument*. The applications of the Internet are determined by what's connected to it, not by how it's constructed. An old-fashioned dial telephone could do only what was programmed into the central phone office switch that the phone was wired to. A computer connected to the Internet to make a call could itself be programmed to use video or to join a chat room. This principle of moving applications to the edge of the network isn't obvious, and was in fact controversial at the time. The Internet is less effective at doing any one thing than is something designed for that purpose, but the Internet is good enough to do almost anything.

The analogy between the architecture of machines and that of computer software or networks starts with the observation that a 3D printer is optimized to move the extruder as quickly as possible, whereas an NC mill must be stiffer to withstand the cutting forces on the spindle. But both types require a motion system to move some sort of head and a way to interpret commands sent to that head. Using the same motion and control system for both and just changing the head would let one machine do the work of both and would be good enough for all but the most demanding applications. This thought led Nadya and another student in the machine-building class, Ilan Moyer (who went on to found another machine company, Shaper Tools), in 2012 to make the PopFab, a rapid-prototyping machine that folds into a briefcase and has interchangeable heads for 3D printing, NC milling, and vinyl cutting. The PopFab is like the laptop of fabrication.

This architectural analogy becomes explicit once the intelligence in the machine is included. Each of the types of rapid-prototyping machines specifies a language to tell it what to do. NC mills typically take G-codes, an odd ancient format that dates back to early computerized photo plotters and fabric cutters. The most common format for laser cutters is HPGL, the graphics language that Hewlett Packard used to talk to its now-obsolete pen plotters. A computer in the machine then converts these commands into instructions for each of its components, such as coordinating the motion of its motors. All of this configuration specification makes it hard to change anything. If you want to add a motor to rotate a part while cutting or printing it, the motor is like an old-fashioned telephone that can't do anything by itself—you have to change the language sent to the machine, and the interpreter of that language in the machine, before the motor can move. An endless series of committees has tried to remedy this problem by coming up with a new universal language for manufacturing. It has been a hopeless task, given the range of what people want to make and how they want to do it.

Around the time we were failing at having fab labs replicate their existing machines, I was doing early work on what became known as the Internet of Things. I was working with a student, Raffi Krikorian (who went on to build and run Twitter's computing infrastructure), and one of the architects of the Internet, Danny Cohen. We showed that the protocols used by servers connected to the Internet could be implemented in a chip costing less than a dollar and occupying a few millimeters of silicon, meaning that everyday objects could be connected to the Internet. That way, the connection between a switch and a light, or a temperature sensor and a heater, could be selected by software, rather than being fixed by the wiring in a house. That thought extends to the parts of a rapid-prototyping machine. We connected all the sensors and actuators in our machines to real-time networks, so that application programs could talk to the devices directly rather than through an interpreter (which leads to the same kinds of misunderstandings in machines that it does in diplomacy). New features could be added just by changing what was connected to the machine's network, rather than requiring changes to its controller hardware.

The historical parallel was completed when Nadya and Ilan, along with James Coleman (who became a lead researcher for the Zahner Company, which does rapid prototyping on architectural scales for making things like Frank Gehry's buildings) began building object-oriented hardware. These were modular machines made out of building blocks that each did a physical task like moving an axis or turning a spindle and were simultaneously nodes in a communication network and software objects in a control program. Now machines lost their fixed identity. For a particular application, you might need a machine that can move quickly or with a lot of force, you might want to print or cut, or you might be making a 2D shape or an intricate 3D form. All these tasks could be composed from a common set of combined hardware and software building blocks and then reused for a different purpose. This becomes the rapid prototyping of rapid-prototyping machines.

As an experiment, we tried sending kits of these machine-building modules to students in fab labs and found that they quickly made a range of fabrication machines, ranging from fanciful to useful. Rather than having to repeatedly reinvent solutions to the same common machine-building tasks, we provided them with motors with integrated attachments for driving a machine, networked interfaces to communicate with the motors, and software components to control everything. In this way, they could concentrate on the interesting question of what they wanted their machines to do.

Jens Dyvik, who runs a fab lab in Oslo after spending two years doing a fab lab world tour, took on a challenge of producing these kinds of modular motion components entirely in a fab lab, with parametric designs so that their proportions could be varied. In doing so, he has eliminated almost all the finished parts that need to be purchased to make a machine, reducing it down to just the motors and electronic components.

The path to a million fab labs, therefore, isn't ordering a million copies each of multiple types of machine that each do one thing. It's manufacturing millions of modular components that can be combined to make many different machines. These could be produced locally from modular designs for maximum customization and independence, the modules could be ordered and assembled individually for maximum flexibility and convenience, or designs for particular combinations of them could be frozen and mass-produced for maximum integration and efficiency. The construction of such a machine becomes dynamic rather than static, changing with whether you're making a circuit, a cake, or a couch, varying the range of travel, the degrees of freedom that can move, and the end effectors that perform fabrication operations.

If you look under the hood, or rather inside the case, something similar is what actually happened when computing became personal. When I was a graduate student in physics in the 1980s, computing was still done on minicomputers. Each of these had unique subsystems; there was no notion of interoperability. That's like a fab lab today, where the only integration between the machines is a person carrying work between them. I was the first person in my research lab to own a personal computer; PCs were just becoming available to early-adopter individuals like me but were still far from becoming universal. Unlike minicomputers, these were, and still are, constructed from a standard set of parts. Motherboards come in a few standard forms, with sockets for interchangeable processors. There are standard internal connectors for plugging in various amounts of memory, graphics, and storage. Standard external connectors connect to choices for a keyboard, mouse, and monitor. Enthusiasts still choose each of these separately and assemble them themselves, small computer companies buy the parts from other vendors, bigger computer companies make their own parts, and these components come pre-integrated for the highest-volume markets, but the whole industry grew up around de facto standards for modularity.

Something similar is now happening with modular smartphones. One of the first, Google's Project Ara (named after my student Ara Knaian, who developed a mechanism for programmable mechanical connections), showed that instead of a mass manufacturer deciding what goes into your phone, you could choose from a kit of reconfigurable modules. Depending on your changing needs, you might want extra battery life, a higher-resolution screen, radios that work on more than one carrier, or a better camera.

Digital photography followed a similar path from specificity to generality. My first digital camera was a closed system that was primarily good for annoying my family by running out of power or storage while I took grainy pictures. These kinds of cameras with fixed functions have largely disappeared (along with the lines of business that made them). For routine picture taking, cameras have merged with communication and computation in smartphones. And for more serious photography, the industry has settled on a standard set of camera body types, with mounts for interchangeable families of lenses and interfaces for accessories like types of camera flash.

For personal fabrication, motherboards or camera bodies correspond to the motion systems that are common to anything that makes anything. The motion systems come in different sizes, shapes, and speeds but are otherwise a universal platform. And the parallel to camera lenses or computer monitors are the end effectors that are attached to the motion systems and add or remove material. In the way you buy a computer or a camera, you could purchase all these elements pre-packaged or, for better performance and more versatility, you could purchase them as parts of a system. But unlike buying a computer or camera, you could go one step further and use your purchase to make more of the thing you just bought.

The commoditization of the components inside a PC has had an un-expected impact at the extreme other end of computing, in the giant data centers that are at the heart of the biggest organizations. Jason Taylor, my former student now in charge of Facebook's computing infrastructure, leads the Open Compute Project. The project has defined a tightly inte-grated, modular PC that can be stacked in huge numbers to build the data centers that are displacing traditional mainframes. In *cloud computing*, users can distribute work over as many of these PCs as they need, and the service providers can add capacity in individual increments to meet the demands of users. Organizations are increasingly finding that it's cheaper, more reliable, and more flexible to purchase computing as a service in this way than to purchase computers.

An emerging concept of cloud manufacturing aims to do the same for making things. To build on the personalization of fabrication, it follows these same steps of standardization, commoditization, and modularization. In this way, cloud manufacturing could provide capacity and capabilities in increments that can continuously span from the needs of an individual to those of a corporation. The capacity and capabilities could be shared re-motely or introduced locally so that, say, a growing bakery could assemble customized manufacturing machines to perform repetitive labor-intensive tasks that are bottlenecks in its business, augmenting rather than replac-ing its workers with accessible automation tools.

UNIVERSAL FABRICATION: 1,000,000 TO 1,000,000,000

The third phase in this roadmap is universal fabrication, when the mate-rials as well as the designs move from analog to digital. In the transition from community to personal fabrication, multiple kinds of machines will be merged into one; in the transition from personal to universal fabrica-tion, multiple kinds of processes will be merged into one: the assembly and disassembly of discrete building blocks. This phase mirrors the pe-riod when computing and communications became so cheap and easy to use in smartphones that markets began to become saturated with more phones than people. To become available to billions rather than millions

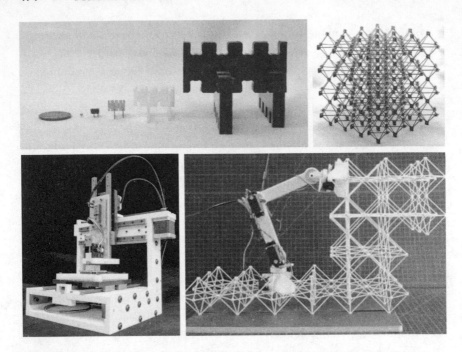

Digital materials and their assemblers. *Clockwise from upper left: Prashant Patil, Kenneth Cheung, Ben Jenett, Will Langford*

of people, the capabilities of a fab lab can no longer require the existing extensive supply chains or generate the same ongoing waste streams. Closing the loop from production to consumption and back again requires that the essential inputs be reduced to a feedstock of a small set of parts and that recycling be replaced with the reversible process of disassembly.

In 2006, struck by the parallel with digital computing and communications, I defined digital materials as those constructed from a discrete set of parts, reversibly assembled with a discrete set of relative positions and orientations. These attributes allow the global geometry to be determined from local constraints. Errors in assembly can be recognized and repaired, and dissimilar materials can be joined. Finally, materials can be disassembled and reused rather than disposed of. The individual elements can be mass-produced; the customization comes in how they're put together. A number of industrial processes—including molding, rolling, stamping, and synthesis—can make large volumes of identical parts at low cost. Lego bricks and amino acids are familiar examples of systems of parts with these attributes.

What's not yet familiar is how many other things can be made by discrete assembly and disassembly rather than by continuous additive and subtractive processes. Among the highest-performance structures made today are the airframes for jumbo jets. To reduce the planes' weight while still retaining their ability to carry tremendous loads, manufacturers now make the airframes by winding filaments of carbon fiber in an epoxy matrix. The latest of the airplanes made this way, Boeing's 787 and Airbus's A350, each required billions of dollars of investment in their supply chains to produce their composite parts. One tool the size of the fuselage or wing is required to lay up tapes of the filament; another even larger tool is required to compress and cure the wound filament in the resin matrix. Then these giant parts must be transported long distances for their final assembly.

In 2008, I was approached about one of the biggest problems in this manufacturing process: how to join composite parts without introducing weakness at the joint. Right now, parts are joined by either drilling holes for bolts or using glue. Although model airplanes have been 3D printed, that doesn't help here for two reasons: 3D printers are nowhere near the size of a jumbo jet, and they can't match the materials performance of what comes out of the composite tools.

In 2012, my student Kenny Cheung (now leading a manufacturing research program at NASA) and I showed that instead of winding one long fiber or extruding lots of short fibers, the solution is to assemble fiber loops that are small compared with the size of the structure, which for airframes corresponds to a few centimeters across. Based on the concept of digital materials, we mass-produced carbon-fiber composite loops and then linked them in cellular structures. These set the world record for the highest-modulus (stiffest) ultralight material, because instead of joining a few big parts, we created sparsely filled volumes that could be assembled in any size or shape. We're now developing the robotic equivalent of ribosomes to place these parts, which work by attaching and moving relative to the structure that they're building, so that a small army of these assemblers, fed a supply of the fiber loops, can build the entire airframe. For this application, it's not necessary to place individual atoms; the fiber loops serve as the basic building blocks. By adding a second part type that can flex rather than be as stiff as possible, we then showed that these structures can be designed to deform in prescribed ways, such as making airplane wings that can continuously change shape like a bird's wings, rather than pivoting the rigid control flaps on an existing wing. Morphing wings have been a long-standing goal in aviation, because they would be more efficient and agile, but prior attempts were hampered by the weight and complexity of the mechanisms required. Because our wings are assembled out of these building blocks, they effectively become the mechanism.

Objects with finer features can be made through the assembly of smaller parts. Although you might expect the discrete construction to nevertheless be noticeable on smooth surfaces, recall that pictures used to be shot on film and video viewed on phosphor screens. From early grainy digital cameras and blocky computer monitors, the resolution of cameras and displays can now exceed the resolution of the eye, so that images are nearly universally recorded and viewed as discrete picture elements, called pixels. Digital materials are built from discrete 3D volume elements, or voxels. Once these shrink to around a millionth of a meter, they can no longer be perceived by our unaided senses.

Beyond rigid and flexible part types to make structures and mechanisms, many more things can be made with just a few more part types. If you zoom in on a computer, you'll first pass through the case. Then you'll see wiring harnesses terminating in connectors on printed circuit boards. The boards have discrete components and integrated circuits attached. The integrated circuits contain myriad elements, which end with the smallest parts of the integrated circuits, with sizes now approaching nanometers, which is a billionth of a meter, corresponding to tens of atoms across. The transition to digital materials doesn't need to start with these smallest features; the components can be replaced in stages.

A typical spacing for wiring and connectors is a tenth or a fifty-thousandth of an inch. Digi-Key, the vendor that I buy electronic components from, stocks a half million connectors. That's not the inventory; it's the number of distinct models, differing in things like the number of positions, how many rows they're in, and which way they face. All of them can be assembled from just two smaller part types, a conducting part and an insulating one. The connectors are then attached to circuit boards, which have conducting traces etched on laminated layers, with a characteristic size of ten-thousandths of an inch. The circuit boards can again be assembled from the conducting and insulating part types, including the tricky-to-manufacture connections called vias that run between the layers.

Also on the circuit boards are discrete components like capacitors, inductors, and resistors. The capacitors contain electric fields and are used in filters and power supplies. Digi-Key stocks around a half million of these also; the capacitors vary in capacity and package size. Again, these can be assembled from just the conducting and insulating part types; likewise for the inductors, which contain magnetic fields. When the part size reaches a micron—a millionth of a meter—it's possible to match the density of capacitance and inductance in components the way they're made today. The resistors, which regulate the flow of current, require one more part type, a single resistive one. By combining resistive, conducting, and insulating parts, you can match the values and packages of—can you guess?—the half million types of resistors that Digi-Key stocks.

We're up to three part types so far to replace millions of components. Transistors for logic require a few more: semiconducting parts whose conductivity can be varied, with versions doped to enhance and deplete the number of electrons. Motors to move things need a few more parts still: permanent magnets, and materials that guide magnetic fields. Taken together, this catalog of parts adds up to tens of types to reproduce the functionality of much of modern technology. That happens to be comparable to the number of amino acids in biology, but these human-made parts have properties that aren't available in biology.

This radical inventory reduction has a number of profound implications. The first consequence comes in the assembler that places the parts, the micro-robotic equivalent of a child's placing a Lego brick. This assembler is finally a truly personal fabricator, replacing the range of machines in a fab lab today with a single integrated process. The second implication is the impact on the supply chain. The assembler requires a feedstock of tens of part types; a fab lab today requires that many vendors to supply all the types of consumables that it needs. The third impact is a result of the reversibility. The existence of trash reflects a lack of information about what it contains. Discrete disassembly is symmetrical with discrete assembly; digital material parts can be reused for many cycles, until the error rate eventually requires them to be remanufactured. This reuse cycle marks the end of practices that fill landfills with obsolete technological trash; now anything discarded becomes instead a supply of parts for new construction. The combination of assembly and disassembly adds up to re-configurability. Rather than make a binary decision to keep or dispose of something, you can continuously modify a product throughout its lifetime to reflect your changing interests and needs.

All these attributes are particularly important at the end of long supply chains. The longest of these is sending things to outer space: material orbiting the earth is roughly worth its weight in gold because of the cost to get it there. But at the end of a mission, satellites either burn up in the atmosphere or become space junk because there's no way to reuse their parts. Making spacecraft that can be disassembled and reconfigured is emerging as one of the early drivers for the adoption of digital materials.

In-situ resource utilization is the term for going into space without luggage, by using locally available materials rather than bringing them from earth. Work on this objective has typically had an implicit assumption that the goal is to pass through the stages of the industrial revolution, to be able to replicate technology on earth. A model for how to do this is a wonderful series of books by David Gingery. The first book is about how to make a charcoal furnace, and the books then progress through hand tools up to making a complete machine shop.

In 2016, I collaborated with 20th Century Fox and NASA on an event that explored the science behind moving to Mars, for the release of the home edition of the movie *The Martian*. The development of digital materials provides a very different answer to the profound question of the minimum requirements to bootstrap a technological civilization. Instead of ending up again with an inventory of half a million types of resistors, capacitors, and connectors, you need to find just tens of material properties (conducting, insulating, magnetic, etc.), form them into discrete building blocks, and then assemble a society.

UBIQUITOUS FABRICATION: 1,000,000,000 TO 1,000,000,000,000

The final phase in the roadmap progresses from digital fabrication for everyone to digital fabrication in everything. This phase corresponds to how the Internet spread from computers to people to things. It will be accomplished by digitizing the construction of the machines as well as the materials. The distinction between a machine and what it makes then disappears, as the materials themselves become programmable.

To see the need for this stage of digital fabrication, consider the speed of the assembly process—a simple calculation shows that there's a serious problem. The fastest current comparable operation is an inkjet printer, which produces on the order of ten thousand drops per second. If a cubic-meter volume were created with cubic-millimeter parts at that ambitious rate (i.e., ten thousand cubic-millimeter parts per second), then the process would take about a day. This is how long it takes a 3D printer to make something that large with features that fine. But if the part size is shrunk to a tenth of a millimeter, the time needed to make a cubic-meter object increases to three years, and if the parts are a hundred times smaller still (a micron, which is around the limit of what our eyes and fingers can detect), the time grows to three million years, a long time to wait for a job to finish.

The process could be sped up by placing more than one part at a time,

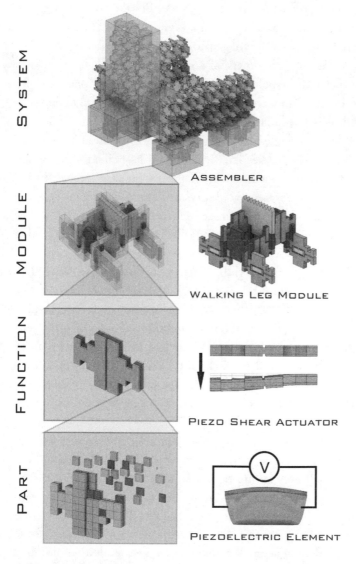

SYSTEM

ASSEMBLER

MODULE

WALKING LEG MODULE

FUNCTION

PIEZO SHEAR ACTUATOR

PART

PIEZOELECTRIC ELEMENT

The hierarchy of assembling an assembler from the parts that it's assembling. *Will Langford, Amanda Ghassaei*

as an inkjet does with its several nozzles. Using more parts at once might speed things up by a factor of ten or a hundred, but placing a million parts at once isn't feasible because of the sensitivity to their relative misalignment. The biological solution to this problem is replication. The ribosomes that make proteins also make ribosomes. The number of ribosomes varies with the needs of the cell but can reach millions per cell. With that many assemblers running in parallel, we're down to thirty years to fill the cubic meter with micron-sized parts—still a long time. But the ribosomes make the proteins to make cells, which then make more cells. The body has around 10^{13} cells; the assembly rate of all those ribosomes in all those cells brings the time down to a tenth of a millisecond to place that many parts. The ribosomes are much slower, though; they run at one part per second (in their case, an amino acid), which brings the time up to one second. And not all cells have that many ribosomes, so assuming a thousand ribosomes per cell, the time to place a billion parts is around a minute.

The point is that what scales isn't how fast an assembler can operate or how many parts it can place at once; it's the exponential growth in capacity that comes from assemblers' ability to make assemblers from the parts that they're assembling. This recursion is essential for the ability to build up from the smallest part size to the largest system size in a sane amount of time.

The technical term for materials that can be instructed to change shape, like the T-1000 liquid-metal man, is *programmable matter*, and the technical term for materials that can autonomously organize, like his droplets, is *self-assembly*. Unfortunately, programmable matter and self-assembly have largely been "desirements," that is, things that exist in research program statements but not in reality. The actual efforts to date have landed far from the vision, with either small numbers of relatively large and complex but capable robotic modules, or larger numbers of simpler elements that can't do very much.

Research on both attributes aspires to emulate the way biological systems can grow, evolve, and repair themselves. But at its heart, biology isn't based on self-assembling programmable matter. Instead, there's a clear division of labor. Instructions arrive at the ribosome via messenger RNA molecules, amino acids arrive via transfer RNA molecules, and the ribosome then follows a carefully choreographed coded sequence to produce proteins. Then, molecules called chaperones guide the folding of the proteins into their final functional form. The individual amino acids aren't themselves programmable; the programmability comes from the boot-strapping of ribosomes assembling amino acids to make ribosomes.

If biology can already do all this, why not just harness it to grow everything? Although it is increasingly feasible to program biology, we rely on many things that can't be made that way. The materials of molecular biology can't synthesize good electrical conductors that can carry high-current and high-speed signals over long distances, or high-temperature components that can withstand the loads and temperatures in a jet engine.

Nevertheless, the same kind of bootstrapping is possible in assembling an assembler out of nonbiological materials. Insulating and flexural parts can make the mechanical structure and mechanisms, conducting and resistive parts can make the wiring and passive components, semiconducting parts can make the logic to program the assembler, and magnetic parts can make the motors to move the assembler. Once one assembler is made, it can then make more assemblers, or anything else out of the parts. The replicator is an assembler that can assemble assemblers out of the parts that it's assembling.

This is the realization of von Neumann's vision of a self-reproducing machine as a model for life. Evolution began with the molecular progenitors of life, progressing from evolving molecules, to evolving cells, to organs, to organisms, to species, and what's now arguably evolving is civilizations and ideas. This final phase in the digital fabrication roadmap completes that arc by allowing evolving bits to arrange atoms, and vice versa.

N

ALAN & JOEL

CHAPTER 6

The Opportunity

Neil has given us a glimpse around the corner into the future, providing a technology roadmap for the next half century of digital fabrication. Like Gordon Moore in 1965, Neil is no Nostradamus; he is simply observing the past decade of accelerations and projecting the likely trajectory moving forward—a trajectory anchored in the underlying science of digital technologies, the active research roadmap, and the growing fab ecosystem.

Having access to this roadmap is analogous to having been in that hypothetical coffee house in 1965, learning about Gordon Moore's observations, and envisioning the projected roadmap for digital computation performance. As with Moore's Law, if Lass' Law holds, digital fabrication performance in the next fifty years will be a billion times greater than it is today. Such a scenario offers the very real potential to democratize manufacturing, transforming how we make (unmake and remake) things, and empower billions of people to make what they consume—creating a more self-sufficient, interconnected, and sustainable society. This is the opportunity.

Realizing this opportunity is another matter. Lass' Law is not written in stone. The pace of acceleration of digital fabrication performance and the impact of these emerging technologies on society will not be determined by some invisible hand. It will be shaped by individuals, organizations, and institutions that recognize the power and promise (as well as the risks) of the technology. Lass' Law depends on people committing to accelerating improvements in capability and reach, as well as to focusing on creating value and mitigating harm throughout society.

We have observed early signs of how fab labs can transform lives, but there is no guarantee the positive potential of digital fabrication technologies will be fully realized. The track record with social systems is clear—they do not easily realign and embrace new technology just because it is possible. Human behavior has to change. Organizations have to change. Old institutions need to adapt and new ecosystems must be forged. Deeply embedded rules and assumptions that guide society and underpin institutions have to be relaxed and refashioned.

Although the people we interviewed for this book have differed on aspects of the roadmap, such as when and how personal fabrication will be widely available in the home, there was broad appreciation for the transformational potential of the technology. The roadmap might or might not unfold precisely as Neil proposes, but there was consensus that digital fabrication capability and reach will improve by many orders of magnitude in the coming decades, with deep implications for society.

Our aim in this final chapter is to offer tangible guidance for ensuring that the arc of the third digital revolution bends toward broadly inclusive individual and community empowerment. In Chapter 2, we examined the threshold challenges of digital fabrication—access, enabling ecosystems, and risk mitigation. In Chapter 4, we put these issues into a social science and historical context, with a focus on rates of change and how digital ecosystems propagate. Now we turn to how to leverage these insights to meet these threshold challenges, co-evolving the social and technical systems.

Throughout our interviews, there was a hunger for future narratives that were neither utopian nor dystopian, but rather aspirational and achievable. As Alvin Toffler highlighted in *Future Shock*, "every society faces not merely a succession of *probable* futures, but an array of *possible* futures and a conflict over *preferable* futures. The management of change is the effort to convert certain possibles into probables, and then to agreed-on preferables."

To assist in the journey toward preferred futures, we begin this chapter with eight aspirational scenarios in which digital fabrication is a key driver for positively transforming society. These scenarios were all co-created with fab pioneers from across five continents. The scenarios address critical social issues such as economic sustainability, literacy and learning, cultural transformation, physical infrastructure, and the very materials out of which everything is constructed. There may not yet be a single integrated vision of true north for digital fabrication, but the combination of these scenarios begins to paint the picture.

To help build the stable steps for urging these and other aspirational futures into reality, we then introduce a model for *predictive transformation*, a framework for helping guide the social and technical systems so they co-evolve to meet collective goals enabled by digital fabrication. We conclude with final thoughts on humanizing the technology so that everyone not only can survive, but also can thrive in the third digital revolution.

HOW TO ENVISION (ALMOST) ANYTHING

It is not hard to imagine a fab lab, even for those who have never been in one. It is more challenging for most people to imagine what personal fabrication looks like in the home. Will it be like a kitchen, a dedicated room with a variety of machine modules and some storage areas for materials? Is there a version that sits on the desktop in a typical office or small business? How does the capability work in rural or village homes, where 70 percent of the global population lives? When it comes to digital materials and programmable matter, it becomes even harder to visualize how a typical person will interact with the technologies.

The four phases of Neil's roadmap get progressively more challenging for the average person to visualize how they might interact with the emerging fab technologies. The first phase is relatively easy; most people can imagine working in a fab lab with digital interfaces to machines like laser cutters and 3D printers. The next phase, which Neil calls personal fabrication powered by "machines making machines" is more challenging. If machines make machines, what are the people doing? In fact, this phase might be better described as the "hobbyist" phase as it involves tech-savvy early adopters being able to take evolving components of a fab lab and configuring them to meet their personal needs. Even in this phase, there are questions, such as where will the personal fabrication equipment reside in a home or other locations. In this case, the connection between personal fabrication and personal computing may not hold as an exact parallel when it comes to everyday behaviors.

With the introduction and spread of digital materials, there is truly a paradigm shift in terms of how people will make, unmake and remake things. How these emerging technologies will impact how we live, learn, work and play is harder to visualize, even for those developing the technologies. What will the digital materials look like? Where will they be stored? How many building blocks will one need to build a chair, a drone, a car, or a house? How easy will it be to dissassemble and reassemble them? In many ways, it is easier to understand the underlying technology principles of the roadmap than to create a mental map for how they will impact daily life.

This is where storytellers and artists, grounded in the science, can play a critical role. Scientists and futurists can look at the research roadmap and make informed predictions about how the future might unfold. Storytellers can look at the same research roadmap and conjure up compelling visions for how the future should unfold. By challenging the world we live in today against the world we want to see emerge, we can galvanize people across distributed ecosystems to transform aspirational futures into reality. Providing a social context and mental maps for the future capabilities of the technology roadmap helps, in turn, to influence and shape the development of the technology.

The aspirational scenarios we present in this chapter are not utopian scenarios. Many tech pioneers present their technologies through rose-colored lenses. The television show *Silicon Valley* satirizes this tendency with its fictional CEOs proudly insisting that they are making the world a better place "through paxos algorithms for consensus protocols" and other similarly absurd, jargon-laced leaps. Techno-utopianism can lull us into a false sense of security that any problems that technology (or society) creates will somehow be solved by technology. Our visions here are grounded in the science roadmap and are ambitious, but achievable.

We must also be careful not to saturate ourselves in dystopian visions of terrifying technological futures. Today's popular movies, video games, novels, and comic books are full of scary visions of technology run amok. When the topic of AI or bioengineering or assemblers assembling assemblers comes up, it's hard not jump to visions of rogue AI threatening humanity or a rogue state (or college frat party) playing with personal biofabrication and changing the human gene pool. There are, of course, very real risks with accelerating technologies, but a constant drumbeat of deeply dystopian scenarios makes many people feel scared and helpless. It risks disengagement or outright opposition.

If we are going to galvanize a generation to help shape the third digital revolution, people need the motivation to lean forward and engage. They need to be inspired. In fact, futurist Alvin Toffler also pointed out in 1970 that "we must vastly widen our conception of possible futures. To the rigorous disciplines of science, we must add the flaming imagination of art."

Cultivating an informed, engaged, and passionate population in a world of increasingly complex and accelerating technologies is one of the great challenges of the twenty-first century. As astrophysicist and science evangelist Carl Sagan famously pointed out, "we live in a society exquisitely dependent on science and technology, in which hardly anyone knows anything about science and technology." To help address this disconnect, we must engage not only pioneering scientists and social scientists, but also storytellers and scholars in the humanities. The eight aspirational futures in this chapter are, we hope, both inviting and plausible. They are all based on the contemporary work of real people, passionate fab pioneers doing amazing work and offering evocative visions for how digital fabrication can improve people's lives. When designing these scenarios, we used a prompt inspired by the world-building process employed by narrative designer Alex McDowell, who has been exploring fab futures since he worked with Greg Lynn in Hollywood to pioneer fab technologies in Hollywood for *Minority Report*, followed by a tour as a visiting artist at the MIT Media Lab.

McDowell has created evocative worlds for movies such as *Fight Club*, *Charlie and the Chocolate Factory*, *Man of Steel*, and numerous other films. He is probably best known, however, for his work designing the 2002 film *Minority Report*, directed by Steven Spielberg. Set in 2054 in Washington, DC, Spielberg wanted the movie to be "future reality," a fictional world built on the foundations of real science. To accomplish this, he pulled together a think tank of pioneering scientists, futurists, designers, and storytellers, including Alex and Neil. The film introduced to a large audience many emerging technologies—from driverless cars to gesture-based interfaces to context-sensitive advertising. Companies like Microsoft, Hewlett Packard, and numerous start-ups have credited the movie for inspiring breakthrough products in the real world.

The *Minority Report* experience inspired McDowell to launch the World Building Institute at the University of Southern California and to create a design studio that combines rigorous science research, expert informed insights with a deeply human lens, and compelling narratives to create new worlds for clients ranging from leading corporations to pioneering social entrepreneurs. McDowell describes this vision in the introduction to *Four*, a collection of short stories published by the software company Autodesk to explore the future of design and technology: "We have the power to build the futures we want to inhabit. Not by following the trends set by our current constraints, but by leading each step forward through imagination and ingenuity. . . . Our future is shared, and storytelling is the common language that allows us to share this vision."

E-Line Media has partnered with McDowell's design studio on a number of science-grounded, expert-informed world building projects which Alan has worked on. The process starts by asking what-if and why-not questions to help open the creative process of imagining the possible, so that the possible can be become the probable and then the reality.

The Fab City movement is a great example of an aspirational vision that is galvanizing and motivating action around the world. The vision asks, "What if our cities could be globally connected, yet locally productive by 2054?" In his Fab City white paper, Tomas Diez captures the aspirational Fab City vision of the future with this declaration (signaled in the introduction): "We need to reinvent our cities and their relationship to people and nature by re-localising production so that cities are generative rather than extractive, restorative rather than destructive, and empowering rather than alienating, where prosperity flourishes, and people have purposeful, meaningful work that they enjoy, that enables them to use their passion and talent. We need to recover the knowledge and capacity on how things are made in our cities, by connecting citizens with the advanced technologies that are transforming our everyday life."

By powerfully articulating a vision of locally productive, globally connected self-sufficient cities, the fab pioneers in Barcelona have created a movement. As we noted in earlier chapters, following the pledge by the mayor and chief architect of Barcelona at the Tenth Annual Fab Lab Conference, over a dozen cities and countries from around the world have signed on.

In each of the following eight new visions of aspirational fab futures, we identify one or more current challenges and introduce an individual whose current work is addressing these issues. We then project forward a few decades to imagine a scenario where the person's work is amplifying and illustrating a potential new reality. We are not making predictions so much as seeking to provide a mental map of possible futures. McDowell calls scenarios like these *provocations* because they are designed to provoke thought and action. Hopefully they will inspire master storytellers across diverse media to incorporate such visions into their art forms.

The first two visions build on the Fab City movement with a focus on self-sufficiency across different geographical locations—distributed networks in rural villages and urban centers. The theme of self-sufficiency, which runs through both scenarios, hints at possible future social and economic models anchored in capability and collaboration, rather than fear and isolation.

What if digital fabrication was distributed across thousands of rural villages?

Most fab labs are in big cities and large towns. There are very few in remote, rural villages. Rural labs have a wide variety of unique challenges, ranging from reliable Internet access and electricity to region-specific issues such as the intrusion of dust from dirt roads into the machines (resulting in down time and maintenance costs) and low literacy levels. Yet if we truly want to strive toward universal fab inclusion, we need to overcome these unique difficulties.

Yogesh Kulkarni, who manages Vigyan Ashram, the prototype for the first fab lab, established in collaboration with MIT, is passionate about solving these problems. The Vigyan Ashram lab, located in a rural region of the Indian state of Maharashtra, educates and empowers local youth with an academic model based on the educational philosophy of Mahatma Gandhi. Gandhi believed that "craft should be the medium of education" and promoted learning curricular subjects through agriculture, weaving, and carpentry by practicing Nai Talim, which translates roughly as "a new education pedagogy." The fab lab at Vigyan Ashram has successfully revitalized this tradition in a modern way with advanced digital fabrication tools, grounded in local culture. Many of Kulkarni's students have gone on to create innovative local solutions to local farming challenges or have started local businesses and have even launched new regional fab labs.

Vigyan Ashram fab lab. *Vishwas Shinde, Vigyan Ashram*

Egg incubator fabricated at Vigyan Ashram fab lab—make almost anything, even chicken!
Vishwas Shinde, Vigyan Ashram

Kulkarni wants to show how this model can ultimately empower all rural communities—the rural equivalent of the Barcelona's Fab City vision. Here is Kulkarni's vision:

▪ *The year is 2038, twenty years after Kulkarni and a network of village leaders and partner fab pioneers across the Maharashtra state announced the "fab village" pledge. They are making great progress toward their goal of bringing basic fab access to all 43,655 villages in the state. Like the Fab City Pledge, the model builds on the phased rollout of digital fabrication capabilities at different scales across the region, but the capabilities are optimized for the unique local rural ecosystem.*

The model began in 2018 when Kulkarni and his team partnered with more than a hundred schools already running the IBT (Introduction to Basic Technology) program he and his team developed. Given that these schools already had basic maker capabilities and, equally important, maker spirit, the IBT schools were a low-friction place to start. The partners worked with the local community to build platforms for collaboration across each of the schools and offered community access to the labs, enabling a combination of private and public funding to help with the initial rollout.

In parallel, Kulkarni and his team began training a cadre of "edupreneurs" (including graduates of the IBT program) to become local fab service providers in the community. These individuals were funded through a new fab-focused microfinance loan program enabling the purchase of single fabrication machines reconfigurable as a laser cutter, a 3D printer, or a 3D scanner. The loans were distributed to ensure that each village has a combination of additive and subtractive capabilities—creating distributed community fab labs. These single-machine enterprises are able to repay their loans through revenue earned by taking in custom design work and renting out time on the machines. All these enterprises helped create a generation of fab evangelists and practitioners necessary for bringing digital fabrication to their communities.

These edupreneurs were also connected to both a local and global knowledge network, mostly accessed through mobile devices. They shared insights on locally relevant projects, from drip irrigation to sanitation to food preservation. They also leveraged the growing materials-knowledge network to get insights from other local fab innovators on how to use the local raw materials and natural polymers from Maharashtra's five agro-climate zones to reduce the cost and waste of the consumables used in fabrication. Over time, some of these individuals have been able to buy additional machines and build larger community-based labs with a greater capacity.

As the number of IBT schools grew and thousands of youth were coming out of the programs more empowered and better prepared for personal and professional self-sustainability, government, industry, and philanthropy took notice. Since Indian law provides for corporations to spend 2 percent of their profits on corporate social responsibility, this growing movement attracted corporate funding because it both produced qualified future employees and created a culture of innovation and made positive social impacts. This new funding stream enabled all the larger villages to launch community fab labs with greater capacity than the local labs.

Initially the fabrication hardware was made by Western companies and imported to India. But soon industrial-sized labs emerged in Maharashtra's regional cities, with the hardware and software increasingly developed locally through globally accessed open-source designs. This local development greatly reduced inefficiencies in the supply chain as well as the overall cost, enabling the exponential growth of all the fab nodes in the regional ecosystem. The national government took notice of all the social, economic, and cultural benefits catalyzed through the regional network. It branded the fab network the Maharashtra Miracle and poured more resources into the community to ensure that all 43,655 villages would have access to basic digital fabrication.

As word of the Maharashtra Miracle fab network spread around the world, other rural regions began to implement and adapt the model. Kulkarni was realizing the vision of Vigyan Ashram's founder, S. S. Kalbag, whose goal in launching the fab lab was "to see India prosper and be a pathfinder to the rest of the world." Kalbag recognized that "this will happen only when everyone can reach his or her own full potential. Hence raise the lowest."

■ ■ ■

What if fab labs enabled self-determination through self-sufficiency?

Often upheld as the poster child for postindustrial collapse, Detroit has faced decades of disinvestment and decay. Less visible is the network of thousands of block clubs, a vital arts scene, and a range of innovative revitalization projects, reflected in the work of Blair Evans.

Evans is an MIT electrical engineering graduate, a serial entrepreneur, a former superintendent of a cluster of charter schools, and the owner of a thirty-acre parcel of land on Detroit's East Side. The parcel is ground zero for a bold experiment in urban transformation. Centered in the Incite Focus fab lab, which operates in the community and in support of K–12 schools and adult life explorers, Evans and his colleagues are boldly exploring new models that "foster the development of impressive, splendid, admirable human beings" and allow everyone to "work and spend less, create and connect more." Evans points out that building self-sufficiency in the inner city is not a new idea, but the impact can be multiplied when

Blair Evans. *Zak Rosen, Can New Work Really Work? (shareable.net),* 2014

fab labs can accelerate the fabrication of everything from food to furniture, as well as personal growth experiences. Let's take Evans's vision into the future.

▪ *Imagine it is 2037 and technological unemployment has accelerated throughout the world, creating great disruption and discord. But Detroit is now a beacon for a "post-salary" future, that is, where individuals can be self-sufficient and find meaning, purpose, and dignity by increasingly making what they consume—self-determination through self-sufficiency. In this model, people spend about thirty hours per week working individually and in community-anchored cooperatives directly making what is needed for the family and community and making additional designs and items for sale and exchange in support of others following similar desires. The remaining fifteen to twenty hours of what was traditionally time on the job is spent combined with other free time to pursue personal passions and personally motivated activities around growth and personal development. This model is now the growing reality in the East Side of Detroit and in urban centers around the world.*

Detroit has also become a pioneer in large-scale permaculture, with well-suited crops such as tomatoes, green peppers, kohlrabi, zucchini, hot peppers, beans, peas, onions, and potatoes. There is sufficient open space for plants and beneficial microbes growing within and above the soil to be a source of energy by converting sunlight, carbon dioxide, and nitrogen into renewable fuel sources. Those combined plantings also provide for a range of accessible ecosystem services such as soil remediation and enrichment, the improvement of air and water quality, and water retention, to name a few. In parallel,

Building permaculture installations in Detroit. *Matthew Piper, "Green City Diaries: Fab Lab and the Language of Nature,"* Model D, 2013

government-sponsored apprenticeship programs offer a path to traditional trades. The combination of the increased food self-sufficiency and increased wage-earning capability has brought stability to the community, as evidenced by formerly abandoned houses being fixed up with innovative fab housing, using designs sourced both globally and locally, but all built locally.

Underneath the change is a fundamental rethinking of the economic model tied to work. In many languages, people repeat what is known as Blair's maxim, which is "selling wholesale and buying retail is a losing strategy." Most people have been working for someone else and, in effect, are selling their labor at wholesale prices, with someone else doing the markup before the final sale. Then, these same people would have to turn around and buy what they need at retail prices. For food, furniture, housing, and many other needs, digital fabrication makes it possible to buy and sell wholesale. There is more barter and exchange than buying and selling, with innovative uses of the blockchain technologies (increasingly sophisticated cryptographic technologies enabling the recording and validating of these person-to-person exchanges).

As Evans puts it, "if living large is being happy, self-sufficient, and part of a community, we can do that." This vision now has traction in cultures all around the world, all of which are emitting and degrading less and enjoying and thriving more.

■ ■ ■

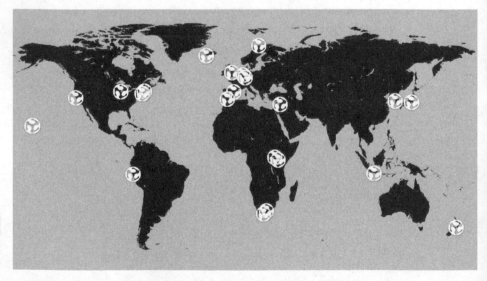

Global map of the fab labs visited by Jens Dyvik in his Jens journey. *Jens Dyvik*

Jens Dyvik presenting laser-cut slippers with the inscribed image of President Barack Obama to the president's paternal grandmother in Kenya. The slippers were designed by a leather craftsman in Japan, then downloaded and customized in Kenya. *Jens Dyvik*

The next two future visions center on learning and mentorship, both qualities essential for cultivating fab literacies. The first vision builds on the personal learning journey of one person who visited more than two dozen fab labs over two years. Learning is also central to the scenario of hands-on project-based learning that begins in Egypt and expands globally. Both scenarios represent methods for turning literacy from a rate limiter to a rate accelerator.

What if tens of thousands of traveling fab mentors circled the globe?

One of the challenges that emerged in virtually every fab pioneer interview and that was reinforced by the feedback from surveys sent to stakeholders is the overwhelming need for experienced mentors throughout the fab ecosystem. The fab pioneers want to mentor those new to the ecosystem, but they often don't have enough time to meet the accelerating demand. Jens Dyvik, mentioned previously, runs a fab lab in Oslo and embodies what could be a new model for personal development and mentorship.

In 2011, after Dyvik graduated from one of Europe's leading design schools, he embarked on a two-year global tour of fab labs, as indicated on the prior page map and above photo. His aim was to find new business models for a sharing-based approach to design. He spent about two or three weeks in each of nearly thirty labs. The full journey cost him the equivalent of around thirty-five thousand US dollars. There were around two hundred labs when he was touring. The trip changed his life in unexpected ways. By the end of the journey, he had implemented the sharing-based approach into his design practice making his designs public and working with an emerging global community of design partners. He also demonstrated how to make a living with the combination of global collaboration and local fabrication. Moreover, he became one of the most sought-after fab mentors. The journey changed him from a student to a teacher. And this brings us to a future vision based on Dyvik's journey which we have called a "Jens Journey":

▪ *The year is 2035. There are now tens of thousands of community fab labs throughout the world. A fixture in nearly all these fab labs are thousands of people on a "Jens journey"—from recent graduates to retirees to the growing number of technologically unemployed to entire families. Like Dyvik's original journey, these travels include visits to multiple labs on every continent. During their two-week visits, the travelers all work as volunteers—sharing what they know and learning from people in each location. The difference is that, in any given week, there might be two or three visitors in each of the tens of thousands of labs. When people travel with their families, which is increasingly common, the learning extends beyond the lab to the host families. Sometimes, of course, with the younger travelers, friendships and romances result in stays that last longer than two weeks.*

A key to making the Jens journeys possible has been the confluence of three enabling factors. First, in 2020, a pioneering Norwegian foundation began awarding scholarships to help people making such a journey. It just started out as benchmarking travel support between labs and slowly expanded into more-extended travel support to visit multiple labs—with labs nominating people who they believe are ready to make such a trip. Concerned that the visits not present a burden on the host labs, there is financial and logistical support for the labs as well as the travelers. The initial support to a lab is prioritized for travelers in the first phases of their Jens journeys—when there is a higher need for mentoring by the host lab. The support for the labs declines as the travelers continue their tours, since the individuals are increasingly bringing mentoring capability rather than needing it. A visitor in the second half of a Jens journey is a welcome addition to any lab.

Around the same time, a growing number of universities around the world began granting credit for one- and two-year journeys if they included a digital

portfolio and met other academic criteria. By 2025, the sources of funding
had expanded substantially to include many private foundations, public agen-
cies, and even commercial businesses seeing this as a way to increase brand
awareness and connect with promising talent. Rather than insisting people
show up at the office every two weeks, unemployment agencies now allow reg-
istration at labs anywhere in the world. Today, Jens journeys have become a
widely valued force for global integration, cross-cultural appreciation, skill
development, and personal renewal.

■ ■ ■

What if students, software, and machines could all improvise together?

Dina El-Zanfaly took Neil's How to Make (almost) Anything class in 2011, as a graduate student at MIT. She had come from Cairo, where she was working as an architect. In her architecture studies in Egypt, the curriculum was highly structured. There was no flexibility to take a class in computer programming, let alone anything like Neil's class.

In 2014, Mohamed Hisham, an IT entrepreneur in El-Zanfaly's home town of Cairo, emailed Neil about setting up a fab lab. Neil connected him with El-Zanfaly, who was already planning to set up a lab. El-Zanfaly and Hisham joined forces and co-founded Fab Lab Egypt. At first, as

Dina El-Zanfaly. Bryce Vickmark, *MIT News*

El-Zanfaly tells it, the lab drew "nerds and geeks, most of whom worked at home." They quickly learned how to use the digital fabrication equipment, but El-Zanfaly noticed that they were not so quick to come up with new ideas on what to fabricate. She set out to develop new approaches to help novices become creative inventors. This goal became the focus of her dissertation, which introduces an I[3] approach to innovation—a process that begins with imitation, followed by iteration, and then improvisation. The approach starts with showing people an example of a finished product and having them figure out how to make their own copy—imitation. This includes figuring out which software and which machines are needed. Then, she has people make one or more guided changes in the item—iteration. After that, people are ready to improvise. El-Zanfaly's vision is for people, software, and hardware to be able to effectively improvise together. Now imagine this vision being realized in schools across Egypt and then around the world:

■ *It is 2030, and there are now hundreds of fab labs across Egypt, some connected to the new Egyptian universities and STEM high schools. Others are located in small and large communities, and a growing number are found in K–12 schools. Initially, the schools were resistant to the concept of fab labs, but when a half dozen pioneering STEM schools began getting great results from their students not only cultivating innovative and passionate learners, but also clearly improving in core subject areas, other schools began to embrace the concept. The fab labs facilitated productive pathways for youth and young adults in the region and, as a result, were valued by government leaders at all levels. Within a decade, having a fab lab in a school was no different from having a library or gym—something that was just assumed a part of well-rounded learning. Teachers came to embrace their shifting role to mentors, coaches, and collaborators instead of primarily lecturers. Schools gradually began shifting from high-stakes testing to innovative portfolio models for assessing and supporting student learning trajectories.*

As schools across Egypt adopted the I[3] model, they were able to share their fab curriculum, professional development, and operating insights with other schools, sharing knowledge and strategies for overcoming barriers to the transition to new hands-on, project-based approaches for teaching and learning. All the tools were also adapted to work well in each of the local communities. English-centric software and documentation, including the more technical vocabulary that often goes untranslated, was all translated into local languages, further reducing friction for adoption and effective use.

To help ensure universal access to all schools, Egyptian school networks developed an innovative program where high school students made digital fabrication machines for middle school students and middle school students made machines that elementary school students used. The older kids mentored the younger kids on both the use of the hardware and software and the application of the I^3 model. They also shared the insights from the Boston South End Technology Center (SETC) fab lab's "learn 2 teach, teach 2 learn" model. Making and mentorship became a constructive focus for youth otherwise facing high levels of unemployment.

At the university and postgraduate level, Egyptian researchers working with like-minded colleagues around the world have begun developing both hardware and software that is more responsive and collaborative to the "fabber" (people in fab labs)—tapping into the broader network of open designs in real time to suggest a diversity of ideas and approaches resulting from the operators' explorations. Throughout the world, the design of software and even hardware has begun to take into account the processes of imitation, iteration, and improvisation, continually adapting and extending El-Zanfaly's vision to align with the local community.

■ ■ ■

Nearly all the feedback we received in our research reiterated the importance of community and collaboration within and across fab labs. The following two scenarios looks at community and collaboration in two very different contexts, one across generations and the other across sectors. In the first scenario, fab labs are revealed as potentially foundational for preserving and extending ancient cultures and traditions. In the second scenario, fab labs connected to multinational corporations are a crucible for learning about collaboration and change that integrates community and industry.

What if advanced technologies preserved and extended ancient cultures?

Personal fabrication is as old as civilization itself. Creating and using tools for self-sustainability and personal expression dates back to over two million years ago. Today, some of the oldest cultures on the planet struggle with how modern society is eroding ancient traditions and cultural heritage. Fab labs could be one more modern intrusion, or as is suggested here, they could support the harmonious bridging of advanced technology and ancient culture.

The Cook Inlet Tribal Council (CITC) is a pioneering Alaska Native social service organization based in Anchorage. A core part of its mission is to build a future where its youth have the ability, confidence, and courage to advance and achieve their goals in a complex, rapidly changing world. Among CITC's initiatives, the council partnered with Alan's company to build a commercial video game designed to share, celebrate, and extend the tribe's culture with a global audience. Sales of the game generate mission-aligned revenue to support CITC's social programs. The game, *Never Alone* (*Kisima Innitchuna*), was developed through an inclusive process with world-class game makers and more than thirty Alaska Native elders, writers, and storytellers. The game is in the Iñupiat language, builds on a story passed down over many generations, and includes twenty-six documentary clips about the culture that get unlocked through gameplay.

The game touched a nerve globally. It has been downloaded by three million-plus players, has been covered widely in the media, and won numerous awards (including a BAFTA—A British Academy Award), and even indirectly led to CITC's launching of a fab lab. Through the E-line partnership, the tribal council leadership met Neil, which led to the launching of the Anchorage fab lab. The lab is run by Renee Fredericks, director of CITC's Youth Education and Employment Services, in partnership with Katie Lee. The lab's mission is to combine cutting-edge educational tools with traditional Alaska Native cultural values and strengths.

Kyla Moore and Elizabeth Paton, both of the Cook Inlet Tribe. *Cook Inlet Tribal Council*

Keyshawn Chickalusion of the Cook Inlet Tribe. *Cook Inlet Tribal Council*

In 2016, Fredericks was recognized at the White House as a Champion of Change, and that was just the beginning. Let's look at the CITC's future scenario.

■ *Flash-forward to the year 2030. The CITC fab lab has now become a network of distributed labs empowering the entire Alaska Native community. Collectively, they have become global pioneers in empowering their youth to mix traditional culture with some of the world's most advanced technologies. Innovation is happening in many areas. In the urban labs, elders are working with young adults who grew up in the lab to mentor the next generation on the bridging of technology and culture. In the remote satellite labs, the community is pioneering how to use the knowledge of local materials to solve local challenges, reducing dependencies on inefficient and expensive supply chains. The village labs are also now collaberating with remote cultures accross the Arctic and throughout the world jointly working on projects and sharing solutions to common problems. In fact, there is now a reverse migration from cities to villages as opportunities for sustainable living are becoming more decentralized.*

The design and operation of a fab lab is also now seamlessly integrated with ancient cultural traditions. The use of a laser cutter to produce scrimshaw is understood as part of a long progression from the use of stones, to metal knives, to electric Dremel tools in carving scrimshaw. Kids who first learned to design and carve scrimshaw on a laser cutter are also interested in mastering the use of older tools. This is not an accident. The design of the fab labs includes workspaces for both modern products and services, as well as traditional crafts and practices. A mix of similar practices are shared with other cultures around the world, from the most remote parts of the planet to teeming urban centers.

Beyond being the place where ancient crafts are appreciated and prac-
ticed, fab labs are where stories and traditions are passed down from genera-
tion to generation. It is now decades since the release of Never Alone, *where*
progressive levels of accomplishment were rewarded with stories and insights
by the tribe's elders or community members. Now traditional storytelling is an
everyday occurrence, integrating into the process of making (almost) anything.
Of course, the mentoring process sometimes goes the other way—kids see what
the elders are working on and share a story of their own. Ultimately, what is
being fabricated in these labs is continuity in the culture itself.

■ ■ ■

What if fab labs were the crucible for disassembling (and reassembling) multinational corporations?

Large multinational corporations can see the handwriting on the wall.
Markets and technology are changing at accelerating rates. Armies of internal change agents and external consultants work daily with industry
leaders to design and implement strategies for becoming more agile and
adaptive. Occasionally, the result is a complete restructuring of the corporation. Sometimes it is a dismantling. The third digital revolution changes
the very nature of how things are made and who makes them, forcing many
companies to rethink their roles in this changing world.

Brazil's Heloisa Neves, a professor and leader in digital fabrication
in São Paulo, supports fab labs because they change educational and
training attitudes—a priority around the world. Her focus, however, is on
how fab labs and maker spaces can change mind-sets within industry. "I
would like to see this mind-set in at least two different spaces," she says,
"schools, integrated in the curriculum, and companies, integrated in what
they do." She adds, "It may be easier for companies to be more open than
for the schools." Toward this goal, she founded We Fab, a maker space in
São Paulo that offers consultancy and activities for companies. In this context, she points to a new fab lab that Isvor set up in Betim, Minas Gerais,
Brazil. Isvor, the corporate university of Fiat Chrysler Automobile (FCA)
Latam, is close to the automobile assembly plant and is open to FCA and
the community.

Marcia Naves, the director of the Isvor fab lab, and Paulo Matos and
Carolina Marini, the fab managers of Isvor, point out that Isvor has one of
the most modern and complete fab labs in South America. But the shiny

Fiat Chrysler Automobile Fab Lab leaders. *We Fab*

Fiat Chrysler Automobile Fab Lab dialogue. *We Fab*

equipment is not what constitutes the heart of the mission of this lab. Although groups of FCA Latam employees and people from the community are working together on digital fabrication projects, they are learning more than the principles of digital fabrication. They are building collaborative skills, developing insights into the interaction of people with new technology, and thinking in fresh ways about the future for the company and community.

The Isvor fab lab is what social scientists call a *boundary object*—it sits at the boundary of different communities and serves in a bridging role. It enables connections across the boundary that would otherwise not happen. The lab is building in the DNA of what could lead to a compelling and aspirational future for the company and the community.

▪ *It is 2030, over a dozen years after the Isvor fab lab was launched, and the lab has become a prototype for culture change in many industries all around the world. In each case, they have adopted the 2016 Isvor model of developing in each person the capacity to adopt any of four personas: the explorer (pioneering new frontiers), the maker (harnessing technology for continuous improvement), the hacker (breaking things apart and finding radically different alternatives), and the networker (connecting the people and ideas as needed). Developing talent along these dimensions, rather than around traditional jobs, has been the beginning of new ways for corporations to function. Neves was correct in her judgment that industry could move faster than could educational institutions in embracing new, highly disruptive ways of thinking.*

Fab labs are now seen as essential crucibles where these personas are forged, strengthened, and introduced back into corporate cultures. By 2030, across diverse sectors, people now appreciate that a person operating as a hacker can be both destructive and generative. They understand that people need the time and space to be explorers and that the whole workforce can be makers, driving continuous improvement. When people adopt the networker persona, they are highly valued. Others recognize that networking is essential connective tissue in organizations that are increasingly organic—almost biological—in how they evolve and change. With these four personas, a corporation can be repeatedly disassembled and reassembled as markets change.

For FCA and many other corporations, the new mind-sets had to come quickly. Distinctions between who is a customer, a manufacturer, or a supplier have been blurring to ever-greater degrees in the years leading up to 2030. Coming to see every member of the organization as able to be an explorer, a maker, a hacker, and a networker helps everyone embrace the increasing ambiguity around the roles of customer, manufacturer, and supplier. FCA is as much a service company as it is a producer of cars. Today, hundreds of FCA fab labs around the world serve as crucibles for the new thinking needed to

be a new kind of a corporation. Thousands of other firms are emulating the model. Back in 2017, Neves said, "My passion now is to transform industry." In 2030, she can look back and see that her early vision of leveraging the technologies and ethos of the fab labs is really transforming industry.

■ ■ ■

The final two scenarios are at the frontiers of material science, with one scenario central to the first two stages of Neil's roadmap and the second centered on the last two stages. In the first scenario, a platform for sharing information on the properties of local materials addresses a key rate limiter for community and personal fabrication—the availability, cost, and environmental impacts of consumable materials used in the digital fabrication process. The second scenario takes us into the future of digital materials and programmable matter. Motivated by the requirements of space exploration, NASA research has the potential to transform how we sustain infrastructure—roads, bridges, buildings, and the like—here on earth.

What if material science became viral?

Today, fab labs use digital technologies for design and fabrication, but depend on analog materials as the consumables. These materials are often expensive, and they need to be extracted, packaged, and shipped. Many are not environmentally friendly and end up as waste. Although advanced digital materials may one day address this problem, they are still decades away for everyday use. We must learn to fundamentally rethink the physical ingredients that we use for fabrication—wood, cardboard, plastic, metal, and more—or we may well find that exponential growth in digital fabrication will create untenable stress on our environment over the next decade.

Alysia Garmulewicz is an assistant professor of circular economy innovation at the Facultad de Administración y Economía, Universidad de Santiago, in Chile. She earned her doctorate from the Saïd Business School at Oxford University, where she studied 3D printing in commons environments. Her commitment is to advancing a circular economy, where resources can be continually reused, as opposed to a linear economy, where resources are extracted, used, and discarded. She begins with first principles: "We need to understand the materials that are around us." Garmulewicz points to nature, where organisms source local ingredients and build complex objects from a nano to a macro scale and continue to

Alysia Garmulewicz. *Saïd Business School, University of Oxford*

reuse of all the resources. Forests don't create trash. Today, she is launching an open-materials database, harnessing the power of distributed knowledge and digital fabrication so that "anyone, anywhere, can make high-performance products using digital designs and the materials around them."

∎ *It is now 2025, and Garmulewicz's open platform, Materiome, is integral to the operation of tens of thousands of fab labs around the world. It has become a combined database, social network, and sharable knowledge resource for understanding, sourcing, and using local resources in digital fabrication. Her open platform challenges the model of purchasing plastic pellets to melt down, metal powder to reconstitute, or other materials through an inefficient, high-carbon-footprint, global supply chain. Instead, enterprising fabricators and makers are using their local environments in new ways. Sand, plant fiber, natural resins, salt water, and other local materials are being gathered, and their properties and uses documented. Mobile, augmented-reality technologies have made it possible to create a local geotagged "augmented layer" of the biological and chemical properties and potential uses of locally available consumables. People can literally walk around their communities and, with the augmented technologies, see trash or desert as accessible natural resources.*

With fab labs able to structure materials from nano to macro scales, abundant natural materials take on high-performance qualities and even outcompete closely guarded chemical recipes. Rather than patenting new compounds, communities and companies alike begin to invest in novel ways to assemble simple ingredients into (almost) anything. In the process, appreciation and understanding of the circular economy is expanding exponentially.

Around the world, people understand that the circular economy is all about the control of energy, materials, and information. Instead of trying to get corporations to track and trace the products they produce, citizens now have the tools to easily measure the composition and other characteristics of the materials they are using locally as well as those from around the world. The Materiome platform enables mass participation in the production of materials data. Such a capability gives citizens a powerful lever of change for the circular economy.

The advances in Materiome have occurred hand in hand with the sharing and reuse of physical samples in geology and the biological sciences. Scientists in these related fields have all encountered the same challenges around what metadata to attach to a given material or physical sample. They all need digital platforms that support distributed input from thousands of sources and that enable efficient search and discovery. Most importantly, they all share a commitment to the open sharing of information. To the biologist and geologist, the advances in fab labs are no surprise—there has long been a tradition of citizen scientists in both domains. In fact, an intergenerational social ecosystem has emerged where biological and geological experts are engaging with teen and young-adult citizen scientists to document through social video and virtual reality all sorts of fun and imaginative fab projects. Material science has become viral, validated, and social.

At first, commercial businesses didn't notice what was going on with Materiome—the platform just seemed to be some fringe activity. After all, fab labs represented a tiny percentage of sales for any producer of consumable materials. The first commercial enterprises to take notice were manufacturers in aerospace, automobiles, and other sectors using advanced materials. For these firms, the properties of materials represent closely guarded, competitive secrets. Thus, many in these industries were surprised to see the open-source platform freely sharing information that they thought was uniquely theirs. They were also surprised to find materials in use that they had never encountered before.

Of course, for the producers of chemicals and other commercial materials, Materiome is a competitive threat. They face existential decisions around how to operate in a world of open information and locally sourced materials. For proponents of a circular economy, the open-source platform is a landmark accomplishment—increasingly, people are using local materials to make what they need with little waste and minimal transportation costs. The adage in politics is "Follow the money"; in digital fabrication, it is increasingly "Follow the materials."

■ ■ ■

What if transportation infrastructure maintained and updated itself?

Massive investments in the transportation infrastructure—roads, bridges, railway lines, buildings, airports—are needed in societies around the world. Bridges built with an expected service life of fifty years are in use today, seventy-five or a hundred years after they were built. Airports are operating at two or three times their designed capacity. The industrial revolution grew alongside the expanding transportation infrastructure, which has taken on ever-increasing importance with globalization. As new transportation infrastructure gets built, will it be designed for a world of growing populations, accelerating technology, changing climate, extensive human migration, and unanticipated societal developments?

NASA's Kenny Cheung sees in digital materials a transformational solution to the problem of aging infrastructure that is performing beyond its designed capability and expected service life. Cheung earned his PhD from MIT, focusing on the frontiers of material science at CBA. At NASA, his work on digital materials and the underlying algorithms has far-reaching implications for structures in space. After all, it would be dangerous and expensive to bring human construction teams into space. Back here on earth, the implications may be just as extensive. He is used to looking deep into the future, so let's do just that.

■ *It is 2040, and we now understand that the physical masses of bridges, buildings, aircraft, and other parts of the societal infrastructure can constantly renew themselves to meet ever-changing service needs. Instead of a service life of fifty years for a bridge, digital materials and supporting robots can be engaged in a constant process of inspection, repair, and reconfiguration, all while the bridge remains in active service. Over time, the bridge may look the same, but it will be an entirely new bridge because of the continuous renovation. The bridge can be redesigned over time to have more or less capacity, depending on changing societal and environmental needs. Instead of trying to reprogram people to be more farsighted, we have reprogrammed the materials to sustain themselves, with people guiding the process.*

In this vision of the future, the supply chain is regional and focuses on the development of an array of modular materials—much as nuts and bolts were produced in 2017. The actual materials are not yet the fully programmable matter of Neil's future (though they are now emerging from research

Digital material structures. *Bridge: Benjamin Jenett, Daniel Cellucci, Christine Gregg, Kenneth Cheung; wing: Benjamin Jenett, Sam Calisch, Daniel Cellucci, Nick Cramer, Neil Gershenfeld, Sean Swei, and Kenneth Cheung*

labs around the world). *Rather, they are made out of the same materials that made up the infrastructure of 2017. Only now they are modular and reconfigurable; they are building-block components for scalable manufacturing systems and are suitable for digital design and assembly. The same materials are used—such as steel in a bridge. But instead of ordering thousands of unique shapes of steel, each with unique drillings and special delivery to the building site, you can now create many modular components as the primary product when the raw material is first processed. There is no need for the reprocessing*

Kenneth Cheung. *NASA*

involved in previous mass production of finished products—people can work with robotic assemblers to make (and remake) custom versions of what they need, by reconfiguring assemblies of these building blocks.

At first, the change was almost imperceptible. It was analogous to the change from physical tollbooths on the turnpike to automated sensors collecting the tolls. In a similar way, people began experimenting in fab labs with reconfiguring materials for personal applications. In parallel, industrial providers began identifying raw materials that could be forged into modular, programmable components for targeted applications. The most advanced applications were in space, where the need for configurable components was the greatest—the cost of getting new raw materials and manufacturing or construction teams into space was just too high.

This modular approach to transportation infrastructure created a massive transportation dividend in public cost savings. Since forward-thinking policy leaders recognized that this smart, modular infrastructure was coming, they also proactively planned for the worker displacement that followed such automation. Getting ahead of the curve, they laid the foundation for various versions of the Fab City Pledge, the "fab village" pledge, and Blair's maxim to ease the transition to an economy with fewer jobs, but higher quality of life and dignity. That is, an economy not as dependent on paid work. Infrastructure self-sufficiency is integral to societal self-sufficiency.

■ ■ ■

These eight scenarios can interweave in complementary ways, building steadily toward a world where people and communities are globally connected and locally self-sufficient. If we want to see these aspirational scenarios converted into reality, we need to begin laying the foundation and building the stable steps to make them happen now. As Gonzalo Rey, who was Moog's chief technology officer, says, "you can always look at visions of the future and see many obstacles. But you can also see possible futures and focus on breaking down the obstacles. Most of the obstacles will be social, not technical." He then points out that "Neil makes these crazy predictions, but over time they seem to come true. It may not be in the exact form he describes, but it will be in that direction." The aim, then, is to tip the balance toward these aspirational, but achievable futures. For that, we introduce a guiding model.

PREDICTIVE TRANSFORMATION

To realize any of these aspirational scenarios, let alone combinations of them, we need to address the threshold challenges identified in Chapter 2. We must promote widespread access to digital fabrication hardware, software, and materials. We need to cultivate the skills and literacies so everyone can effectively leverage these technologies. We need to create an enabling ecosystem to accelerate shared objectives. And we need to ensure the ongoing mitigation of risk.

In addition to addressing the threshold challenges, there are enduring tensions that, even if they cannot be completely resolved, must be continually managed. Among the various digital fabrication stakeholders and communities—fab labs, maker spaces, hacker spaces, and so forth—there is a tension between independence and interdependence. Hardware and software developers will need to balance ease of use with making their technology open and extensible. Governance will always struggle between balancing the efficiency of centralization and hierarchy with decentralization and distributed agency. Mentors face a persistent choice between their commitment to teach and their desire to advance their own projects. There will be ongoing tensions between emergent ecosystems and entrenched institutions.

As we sought to provide guidance on shaping the third digital revolution—with these aspirational scenarios, threshold challenges, and enduring tensions in mind—we couldn't find a model that was well matched to the task. In particular, we needed a model that was optimized for both digital and physical worlds, had a complex mix of distributed stakeholders and interests, and didn't depend on a single governing person or entity directing its progress. Through the process of writing this book, we developed such a model that we call *predictive transformation.*

The model is predictive in that it is anchored by a projected technology roadmap, based on observable indicators of accelerating digital fabrication capability. It is transformational in that we are not interested in simply projecting, but also in shaping how the technology can co-evolve with the social systems to cultivate positive outcomes for society. We have written this chapter more for the path creators than the path observers.

The model builds on and contrasts with many established models of change. Most of the early models, from the dialectic model attributed by Hegel to Kant (thesis, antithesis, synthesis) through to Kurt Lewin's influential 1947 change model (unfreezing, changing, refreezing), anticipate transformation and the need for change. But they also assume a linear pace of change, concluding with a new steady state rather than continuous and accelerating change.

Current popular organizational-change models also conclude with a new steady state rather than an assumption of continual change. For example, John Paul Kotter's 1995 highly cited model for leading change presents eight steps for managing change and concludes with institutionalizing the change (a new steady state). Similarly, William Bridges's 2000 model for managing transitions is helpful at the individual level in documenting how people need to let go of the old before they can embrace the new. Yet, letting go of the old becomes an unending state in a world of exponential technologies. The closest fit to our model is W. Edwards Deming's model for continuous change, which is a cycle of plan, do, check, and adjust, although his model was not designed for harnessing the power of distributed ecosystems and broad societal change.

A number of popular models for analyzing patterns or cycles of new technology adoption or impact include Everett Rogers's 1962 "diffusions of innovation," Geoffrey Moore's 1991 "crossing the chasm," and Clayton

Christensen's 1997 "disruptive innovation." The US government has also developed several helpful rubrics, including NASA's "technology readiness levels" and the Department of Defense's "technology readiness assessment" to assess the maturity level of a particular technology. Although all these models have valuable insights for introducing technologies into markets or channels, none combines the analysis of continuously accelerating technology with mechanisms to engage and empower diverse, independent stakeholders around aligned social-impact objectives.

We designed the predictive transformation model for the third digital revolution, but the framework could be relevant for identifying and shaping other accelerating technologies. Of course, this proposed model is just one of many possible ways to organize thought and action going forward in a world of accelerating technologies. The model is structured around the following four phases and activities within each phase:

Anticipate: Rates of Change
Align: Stakeholders
Cultivate: Enabling Ecosystems
Co-evolve: Technology and Society

Since the model requires a clear understanding of accelerating technologies, it assumes continual cross-sector collaboration among scientists,

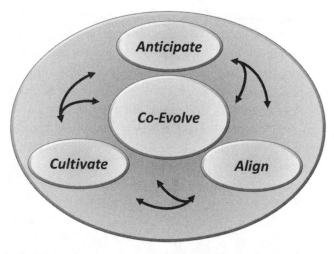

Predictive transformation model as a continuous process. *Joel Cutcher-Gershenfeld and Alan Gershenfeld*

technologists, and social scientists, as well as storytellers, policy makers, facilitators, and others. Like all models, it is not a recipe; rather, it is designed as a framework to help inform the process and suggest enabling platforms, tools, and practices that can be adopted, adapted, and extrended by distributed stakeholders. The model should be flexible, adaptable, and extensible to work in a variety of contexts at local, regional, national, and international levels.

Although the model has four elements that are presented here in sequence, the interactions are dynamic. The figure on the previous page suggests that the first three parts of the model—observing rates of change, aligning stakeholders, and cultivating emergent ecosystems—are relatively sequential, but when each phase makes progress, it contributes to the co-evolution of technology and society. Furthermore, it is a continuous cycle—a spiral that continues upward as long as there are stabilizing efforts—scaffolding—to hold the gains all along the way.

Since a book is, for the most part, linear in presentation, we address each phase of the model sequentially. In the final section, on co-evolution, we can most clearly see the synthesis.

Anticipate: Rates of Change

All the aspirational scenarios we described, and many others that are emerging, are built on an assumption of continued exponential improvement of digital fabrication performance. Thinking in exponential terms rather than linear is not instinctive. Thus, we begin our model with the need to anticipate the rates of change of digital fabrication technologies. This requires a basic understanding of the underlying science driving the accelerations, as well as the active research roadmap. When co-evolution pioneer Stewart Brand was asked, "What do you consider the most interesting recent [scientific] news?" in a 2016 online salon query from agent and thought leader John Brockman, Brand's response was instructive: "Science is the only news. When you scan through a newspaper or magazine, all the human-interest stuff is the same old he-said-she-said, the politics and economics the same sorry cyclic dramas, the fashions a pathetic illusion of newness, and even the technology is predictable if you know the science. Human nature doesn't change much; science does, and the change accrues, altering the world irreversibly. We now live in a world in which the rate of change is the biggest change. Science has thus become a big story."

Brand says that the technology can become predictable if you know the science. In our model, we urge an initial focus on understanding the science and how it provides the framework for the research roadmap as well as opportunities and risks for society.

In our survey of leaders in digital fabrication, we asked about the importance of understanding the underlying science of digital communication, computation, and fabrication. The responses suggest that for over 70 percent of the stakeholders, an understanding of the underlying science is very important. At the same time, nearly 40 percent said that it was hard to do. These responses are illustrated with the same visualizations that were used in Chapter 2 (with the middle representing the central tendency and the outliers on the outside).

As the figure indicates, there is a wide variety of responses on difficulty. Improvement is difficult when there is a great deal of variability in a system. For those wanting to understand and thoughtfully debate rates of change, Neil's chapters are required reading. That doesn't mean you have to agree with all of his assumptions, but it is an informed starting point for discussion and debate. At present only about a quarter of the fab leaders report that it is very easy to understand the underlying science. Throughout the writing of this book, we continually pushed Neil to make the more scientific or technological-focused sections more accessible both for us and for the readers as the material he covers is necessary to understand likely rates of change. The more we understood the science and

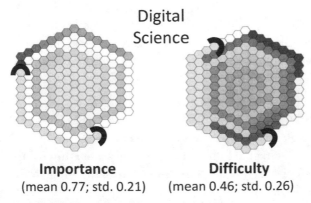

Digital Science

Importance
(mean 0.77; std. 0.21)

Difficulty
(mean 0.46; std. 0.26)

My understanding of digital science (.31 gap between the two means).

Importance and difficulty in understanding the underlying science of digital communication, computation, and fabrication. *WayMark Analytics*

technology roadmap, the more we understood the critical steps in shaping the third digital revolution. The shared understanding of the underlying science also led to the codification of Lass' Law—which helped organize our thinking. This also helped ensure that we weren't talking past each other when discussing digital fabrication performance, making our collaboration more effective.

Interrogating Lass' Law

The rates of change in the digital fabrication roadmap may accelerate or decelerate over time, much as, in 1975, Gordon Moore revised his projection from digital computing performance doubling every year to every two years. Still, the core definition of digital computation performance remained consistent for Moore and that needs to be true for digital fabrication as well—so that discussions and debates on rates of change are using the same measuring stick. For Lass' Law, as discussed throughout the book, we use this definition for digital fabrication performance:

> The equivalent capability of today's fab labs to make (almost) anything, fabricating physical forms and programming their functions.

Agreeing on the definition of digital fabrication performance is essential to enable diverse stakeholders to effectively discuss, debate, anticipate, and shape the social implications of the projected roadmap and rate of change. There is a cottage industry of futurists, pundits, and journalists who make projections about the future of technology and society. Most of these observers assume that the future of digital fabrication is simply better, faster, cheaper 3D printers. Their conclusion is a good indication that they have not meaningfully immersed themselves in the technology. Unfortunately, this lack of awareness is also a problem with many educators, policy makers, philanthropists, and investors. As we have shown throughout this book, 3D printing is an important additive process for digital fabrication, but it is only one tool within the broader capabilities of a fab lab. As Neil points out, seeing the future of digital fabrication through the sole lens of 3D printing is like seeing the future of cooking through the sole lens of microwave ovens.

Given the complexity of emerging technologies, many people are understandably hesitant to look around the corner into the future and make predictions about technology trajectories that can inform action. We certainly see this reluctance in many social scientists and policy makers.

Indeed, it is commonplace in academic presentations, particularly by junior scholars, to feature the quote "Making predictions is hard, particularly about the future," which is variously attributed to Niels Bohr, Samuel Goldwyn, Yogi Berra, Mark Twain, Nostradamus, and others. In the context of an academic talk it might be seen as a form of self-deprecating humor. Yet it has the subtle, but unmistakable impact of reinforcing a bias toward making small incremental predictions and avoiding anything that might be seen as standing out from the crowd.

Even some of the most sophisticated thinkers and practitioners from the first two digital revolutions take a cautious stance. In their recent book *Whiplash: How to Survive Our Faster Future*, Neil's colleague Joi Ito, director of the MIT Media Lab, and coauthor Jeff Howe, a contributing editor to *Wired* magazine, strongly assert that no one can predict the future and it's a fool's errand to try. Together they then recommend nine organizing principles for navigating exponential change. Among the recommendations, they suggest prioritizing emergence over authority, pull over push, risk over safety, and systems over objects. All the principles are designed to address the observation that "our current cognitive toolset leaves us ill-equipped to comprehend the profound implications posed by rapid advances in everything from communication to warfare."

Although their nine principles are full of great insights into new mindsets that foster agency and collaboration in a digital age, we disagree with Ito and Howe's assertion that the pace of change is moving so fast that prediction is not advised. They make this point to foster self-reliance and flexibility—the importance, as they put it, of having a compass rather than a map. In contrast, we believe that you cannot foster the needed agency without roadmaps—so that distributed individuals can be empowered to collectively debate, discuss, and ultimately help accelerate, decelerate, or otherwise change the roadmap. You need both a compass and a map. A more forward-thinking attitude can enable observations to go from aspirational projections, to collective action (using many of Ito and Howe's principles), to help to realize these projections. As management sage Peter Drucker said (along with many others also credited with the comment), "the best way to predict the future is to create it." In this sense, anticipating rates of change is a team sport and a hands-on activity.

Gordon Moore was very precise in his original observation of digital computing performance. He observed that the number of transistors per square inch on integrated circuits was doubling every year, and he recognized the implications if this rate of change continued. Over time, however, Moore's Law came to represent many interrelated technologies that were enabling computing capability to become exponentially better, faster, and cheaper and thereby transform society.

For digital fabrication, Lass' Law projects that digital fabrication performance will double every year and a half, in light of past indicators and the current research roadmap. Whether this rate of change holds is very much up to all the stakeholders who, literally, have a stake in this exponential rate of change. If, like Moore's Law, Lass' Law evolves from an observation to a benchmark to a collective objective across diverse stakeholders, then, as we noted in Chapter 4, we may indeed see digital fabrication performance increase at a rate similar to Moore's Law for the next half century.

With observable data, a clear definition, and informed projections in hand, we can anticipate and envision what the third revolution might look like. We collaborated with fab leaders to draft the scenarios opening this chapter, beginning with current data considering what would happen if things were to unfold in transformative ways. On the analytic side, we have drawn a graph that corresponds to the four stages in Neil's roadmap. This graph illustrates how each successive technology will grow at a faster rate of change and, when it catches up with the preceding technology, the preceding one is likely to level off, forming the sigmoid curves the Neil mentioned in Chapter 3. The exact trajectories can't be precisely specified, and the graph is not to scale (we could not both make it fit on a page and indicate the shape of the first two curves). Our point is to illustrate and thus anticipate the reality of successive waves of accelerating change.

This figure signals the scale and scope of the social challenge. Individuals, organizations, and institutions face a challenge in keeping pace with the rate of community fabrication, because they must ensure the success of tens of thousands, or hundreds of thousands, of community fab labs. The task gets more difficult when we try to keep pace with the next curve, personal fabrication, and need to enable millions of early adopters to succeed with their own personal fabrication technology. The task becomes breathtaking when we are matching social systems to the final two curves, where we are operating on the scale of billions of individuals or trillions of things.

Approaching the challenge of anticipating rates of change analytically is instructive, but only goes so far. This is where the diverse disciplines come in—providing alternative perspectives and future visions that are grounded in the data. The data and its visualization help make visible the probable futures and lay the foundation for aligning stakeholders—path

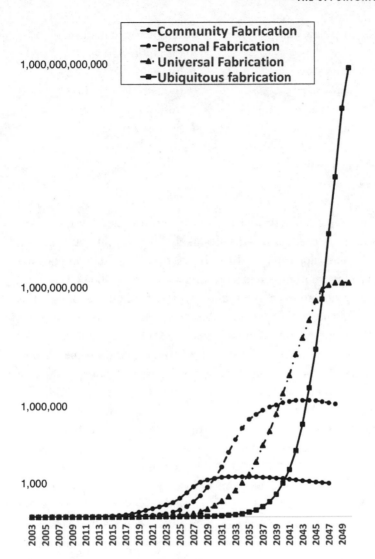

The four stages of the third digital revolution, projected over time. *Joel Cutcher-Gershenfeld*

observers with path creators (technical and social)—to come together and co-create preferable futures.

Align: Stakeholders

Throughout the book, we have highlighted a rich tapestry of stakeholders actively engaged in the fab community and broader maker movement, including:

- Fabbers, makers, and hackers
- Hardware providers
- Software providers
- Materials providers
- Researchers
- Educators
- Government policy makers
- Community organizers
- Philanthropists
- Investors
- Industry leaders and workers
- Service providers
- Media leaders
- Families

In working toward stakeholder alignment, even specifying the stakeholder categories is instructive—most stakeholders operate with mental maps that only encompass parts of the list. So stepping back to consider the full list is a first step. A second step involves a deeper appreciation that none of these groups are monolithic and many individuals have identities tied to more than one of these groups. Moreover, the interests—what some would call the value propositions with respect to digital fabrication—for these stakeholders are both common and competing. And all these interests are dynamic. In such a context, progress depends on having sufficient alignment among stakeholders to enable collective action. Invariably, the initial alignment will be among just some of the stakeholders on some of the interests. As we will see shortly, these small agreements can then build toward broader agreements.

Joel's research team defines stakeholder alignment as "the process by which independent but interdependent individuals, groups, and organizations orient and connect to advance their separate and shared interests." This approach is different from how most people use the term *stakeholder*. Typically, people talk about "stakeholder management" and "stakeholder engagement." When people say they need to do stakeholder management, they are usually saying that there are parties whose opposition is anticipated and needs to be neutralized. When people say they need to do stakeholder engagement, they are usually saying that there are parties whose support is desired and needs to be included. Both stakeholder management (i.e., dealing with opposition) and stakeholder engagement (i.e., building on enthusiasm) are important, but incomplete. In addition, stakeholder alignment is needed. This dynamic process assumes that there will be a mix of common and competing interests, requiring a continuing process of dialogue, discovery, negotiation, and action. Power differences and other dynamics need to be taken into account, and progress is ultimately achieved when there is sufficient alignment among independent, but inter-dependent, parties to support collective action.

Small Agreements Lead to Grand Bargains

Although there are broad, sector-wide goals held by most of the global fab stakeholders—goals such as ensuring fab access and literacy—there are often specific goals that are shared by a much smaller group of stakehold-ers where there can be short-term gains and momentum building. Progress is most likely when groups of stakeholders accept that most interests are negotiable. Even though some red lines can't be crossed, history shows that interests can change when new options are identified and when cir-cumstances shift. Ultimately, society's institutions are a product of pat-terned behaviors, including the patterned agreements. Coalitions and small agreements among stakeholders—building on dialogue about their respective interests—can add up over time to form *grand bargains* ad-dressing the biggest challenges facing society. Thus, grand bargains build on multiple smaller agreements, which rest, in turn, on the constructive integration of stakeholders and their interests.

As Gregg Behr from the Grable Foundation points out when talking about the emergence of Remake Learning, "much of the collective impact came from many small decisions made across the network." Progress was made through a variety of flexible, small grants that encouraged stake-holders to take risks and experiment together. Over time, this approach led to a great many insights, programs, and initiatives that helped shape the larger vision of reshaping the future of teaching and learning in the Greater Pittsburgh region.

There is also an interesting connection between small agreements leading to larger goals and the intersection of human psychology and game design. Mihaly Csikszentmihalyi, a Hungarian psychologist, developed the concept of *flow*. He found that when a person's skill is too low and the task too hard, people become anxious and frustrated. If, however, the task is too easy and skill too high, people become bored and disengaged. When the skill and difficulty are roughly proportional, people enter a flow state, espe-cially if they are working toward personally meaningful goals. In his book *Good Business: Leadership, Flow, and the Making of Meaning*, Csikszent-mihalyi describes flow as a state of concentration or complete absorption with a given activity and situation—a state in which people are happiest.

Since games are all about giving the player agency to make decisions and providing continual feedback to help them on their path to mastery (in the context of the goals of the game), game designers are particularly good at designing for flow states. They have also learned to design for collective flow, where teams and communities can continually take on small challenges that can lead to larger goals. Often big societal issues, like addressing climate change or ensuring universal fab inclusion, can be overwhelming. It is essential, therefore, to break these large challenges into achievable steps where everyone feels like their actions matter, they are making a difference, and they are part of a large movement making progress toward a collective goal.

Build Trust

Another key element of Remake Learning's success in aligning diverse stakeholders in Pittsburgh was prioritizing time to build trust through informal social interactions. Things started slow, with a series of pancake breakfasts. Over time, each event drew in ever-greater numbers as people reached out through their personal networks and brought in two or three more people who they knew just had to be in the room.

To ensure that community fabrication reaches its full potential, communities, regions, and even entire nations need the equivalent of pancake breakfasts. For example, when a network of fab labs or maker spaces is being rolled out, debates often surface around the terminology to be used: fab labs, maker spaces, hacker spaces, or some all-encompassing umbrella term. The debate is not just about words—the underlying cultures and operating practices are related but distinct in these different domains.

There are also likely to be debates among the potential hosts for community digital fabrication. Should the hosts be in libraries, K–12 schools, community colleges, universities, museums, or free-standing community centers? Will the vision be compromised or advanced if it is an even split among these six possible hosting options? Universities often have great resources but can't always offer open community access. Community centers and libraries have open access but often don't have experience with new and complex technologies. Ultimately, an essential ingredient for setting up and effectively running local labs is adequate time to build trust and clear communication across stakeholders.

In this spirit, when success depends on aligned actions across independent and interdependent stakeholders, it is essential to take the time to build a culture of trust and respect so that everyone is collaboratively solving problems—but each is still respectful of each organization's independent objectives, capabilities, and constraints. In some cases, of course, interests are not just different, but opposed.

The *Never Alone* backstory illustrates building trust across diverse stakeholders with shared interests. Before launching game development, the Cook Inlet Tribal Council and E-Line devoted months to building trust, spending time in each other's worlds, and getting to know each other's families and communities. As partners, both organizations dedicated time to designing deal terms to ensure they were in the same shoes, working toward aligned impact and financial objectives with shared governance. This way, when challenges arose (and many did), the partners could take advantage of their diverse perspectives and skills to solve problems together. The result was not only a successful and evocative video game but also a much deeper partnership that included a portfolio of initiatives including fab-related products and services.

Reduce Variance

As we described in Chapter 2, the fab, maker, hacker, TechShop, and other related communities certainly have shared interests but also varying priorities, methodologies, and community norms. Understanding these variances is important because a fundamental principle of systems change is that you have to reduce variation before you can improve with collective action. The following figure shows a hypothetical dart game that helps illustrate this point.

In this game, the player on the left has the higher score in this round. However, the player on the right is easier to coach (just shift everything over a little). In contrast, if the player on the left were instructed to go up a little or down a little or over a little, the result would just be more variation. By analogy, improving the highly variable social system associated

Hypothetical dart game illustrating the challenge of improving a variable system. *Joel Cutcher-Gershenfeld*

with digital fabrication—even at this early stage—first involves reducing the variation. But you can't even begin to reduce variation if you don't understand the stakeholders, their interests, and how these interests evolve over time. In many of the z-flowers presented in this book, the variance is high (bimodal, in fact). Improvement can't just center on trying to move the means. (Of course, if you don't know where the bull's-eye is, then the approach of the player on the left could make sense—reducing variation must always be balanced with maintaining a diversity of ideas.)

In light of our stakeholder alignment analysis, we do have a few bull's-eyes in mind. Across all the fab stakeholders we interviewed, we found very strong alignment around the importance of addressing two key threshold challenges: universal fab access and literacy. This is reflected in the priority given to open access for fab labs and is clearly highlighted in the fab charter. Achieving universal fab access won't be easy; it will require aligned and sustained efforts across a diverse community of distributed individuals, organizations, and institutions and, importantly, the support of an enabling ecosystem.

Cultivate: Enabling Ecosystems

Once stakeholders understand projected rates of change and are sufficiently aligned around shared goals, such as universal fab access and literacy, the key mechanism for action centers on cultivating enabling ecosystems. Enabling ecosystems can help advance common interests and navigate competing interests, all at exponential rates of change.

In Chapter 2, we identified the need for cultivating an enabling fab ecosystem as a threshold challenge for realizing the ambitious goal of universal fab access and literacy. In particular, we highlighted certain essential elements of such an overall fab ecosystem, including;

- A diverse mix of public, private, and philanthropic financing for fab initiatives;
- Effective collaboration and knowledge sharing within and across fab labs;
- Widely distributed mentorship and leadership for digital fabrication
- Open and robust marketplaces for fab products and services;
- Agile governance mechanisms aligned with the values of the emerging fab community

Without these elements (each of which can be thought of as having its own ecosystem), digital fabrication technologies will not expand in performance and reach consistent with Neil's roadmap. This will significantly reduce our ability to leverage fab technologies to improve society and, simultaneously, increase the likelihood of destabilizing fab divides.

Cultivating an enabling fab ecosystem will require a blend of top-down scaling initiatives as well as middle-out and bottom-up propagation. The top-down scaling will be necessary for large-scale investments in infrastructure and research, while the middle-out and bottom-up propagation will be necessary to engage and empower a diverse mix of global stakeholders, with a complex range of motivations, interests, skills, and operating in very different regional contexts. The top-down models are familiar; what is new here is cultivating ecosystems. The work of cultivating ecosystems is a widely shared opportunity and responsibility. All stakeholders need to think like system architects since their distributed cultivating behaviors are, in effect, creating the ecosystem.

In outlining the overall strategy for cultivating enabling ecosystems, we start with government and large non-governmental institutions (NGOs), who have the capability to launch initiatives beyond the scope of most organizations and individuals (similar to pivotal early investments in the develop of the Internet). Concurrent with the role of these public-interest institutions is the role of the private sector (similar to the early investment of Intel and other pioneering digital companies). Innovators in universities (both social and technical) and social entrepreneurs make up the third leg of ecosystem. The alignment of these stakeholders is not a one-time event, but a continuing accomplishment, advancing progress toward the threshold challenges of universal fab access and literacy (our primary focus here), and other threshold challenges such as risk mitigation.

Government and NGOs

There are certain initiatives that can happen only at a scale and scope provided by governments and large NGOs. While some of these initiatives will need to be built from scratch, many can be built on existing digital access and literacy initiatives. These initiatives have already developed multi-stakeholder partnerships as well as critical infrastructure that could be a significant rate accelerator for the third digital revolution.

A relevant international example is the World Bank's Global Connect initiative, which aims to bring 1.5 billion online by 2020. Significant international human, political, and financial capital, along with cross-sector stakeholder alignment, has gone into developing this initiative. Launching the initiative has required years of trust building among those wanting to ensure that everyone benefits from essential digital technologies. Similarly, President Obama's ConnectED initiative (part of the Federal Communication Commission's E-Rate programs) aimed to connect 99 percent of American students to high-speed broadband in their classrooms by 2018. Real progress has been made toward this goal. Rather than starting from scratch, forward-thinking government and NGO leaders can build on these and other digital access initiatives to accelerate fab access.

There are also fab-specific government initiatives that can be accelerated. For example, the National Fab Lab Network proposed by Rep. Bill Foster and described in Chapter 1. This is a rare example of a truly bipartisan initiative in an often-divided Congress. Since digital fabrication represents an approach to next generation manufacturing where capabilities remain local but knowledge sharing is global, it cuts through the false dichotomy between global connectivity and local self-sufficiency. Making these benefits clear and visible to the public will be essential. When we asked Kumar Garg, who worked in education and technology issues at the White House Office of Science and Technology Policy, about building support for digital and fab access, he expressed the importance of making access connect to people emotionally:

> It is important to talk about everyday realities that will really grab people—such as kids sitting in the parking lot of McDonald's to get Internet access to do their homework because they don't have it at home. Ensuring access is a valuable public investment and it is much easier to build inclusion in by design early on, when the technology is still being shaped. It is much harder to do so later.

As digital fabrication technologies develop, the role of government and leading NGOs will evolve. Today, these institutions can use their unique capabilities (e.g. ability to convene, evangelize, incentivize and, in some cases, fund) to expand the network of community and school fab labs. They can also leverage their ability to catalyze foundational research to help accelerate the later stages of the digital fabrication roadmap where technology is currently the rate limiter.

Philanthropy and Impact Investors

Addressing fab access and literacy will require significant investment from a variety of sources, public, private, and philanthropic. Private investment will follow the demands (or projected demands) of the market. But these investments will not sufficiently advance fab access and fab literacy. Philanthropic and public investments will directly focus on these societal objectives, but are most effective as catalysts rather than as sustaining sources of support. Traditionally, philanthropic and private investment have been seen as two ends of the investment spectrum—philanthropic seeking social impact and private seeking market-based returns. There is now a third element, which is the growing social-impact investment sector that uses market-based mechanisms to make social impact. All of the above will be essential to tackling the threshold challenges of fab access and literacy and are described here in sufficient detail so that they can be effectively navigated by the fab community.

Starting with philanthropy, there is a great opportunity for a leading philanthropy or an emerging family foundation to become the 'Carnegie' of community and school fab labs. As mentioned previously, the Carnegie model of supporting open-access public libraries is a clear analogue to supporting community open-access public fab labs. Although he was often brutal in his business practices, Carnegie was transformational in his philanthropy. Over the span of a few decades, he funded more than twenty-five hundred libraries. Before Carnegie, the concept of the public library was not commonplace—after Carnegie, it became a natural assumption that every major city and town would have a public library, an assumption that has held for more than a century. A forward-thinking philanthropist could accomplish a similar transformation for community and school fab labs.

Forbes estimates that there are more than 1,800 billionaires, with a combined wealth of seven trillion dollars. Over 130 of these billionaires have taken the Gates-Buffet pledge to give most of their wealth away to benefit society. Many of these billionaires made their wealth in the first two digital revolutions and are interested in using new technologies for social good. Many want to support education, community development and entrepreneurship and are looking to make a lasting, positive impact on society. Funding a local, regional, national, or even global network of fab labs meets all these requirements. Local, national and international philanthropy has been critical to the initial ten years of fab lab growth, primarily with a focus on setting up and sustaining individual fab labs. It is now time for more ambitious philanthropic initiatives.

As the number of public community and school fab labs grows, we can learn a further lesson from Carnegie's library initiative. John Leslie King, former dean of the University of Michigan School of Information, reminds us what happened in the early stages of the Carnegie roll-out: "A few years after Carnegie started building the libraries across North America, they were failing in droves. The Carnegie Library Commission asked leading experts to find out why. Their report said it's because there were no librarians, and there were no librarians because there were no library schools! The technology (the materials and buildings) was essential, but by itself was not enough. There must be people who can build and maintain the systems." In many ways, the schools of library science (many of which have rebranded themselves as schools of information) were the enabling platforms of an earlier era. They taught the enabling practices, to provide society with the qualified people to serve as guides, coaches, and mentors.

In addition to private investment in fab products and services and philanthropic investment, there is a third type of investment—social impact investment—that has a crucial sustaining element. Social impact investing operates under various banners: double bottom line (financial, social), triple bottom line (financial, social, environmental), blended value, program-related investing, mission-related investing, and other emerging models. By investing in both financial and social impact, it ensures alignment between investors and entrepreneurs, allowing for co-evolution of the market development and social impact.

Today most social impact investing is focused on community development, the environment and, increasingly, workforce development and education. Although digital fabrication touches all of these impact areas, the sector is not on the radar of most impact investors and many fab social entrepreneurs are not aware of the growing impact investing sector. To help address this knowledge, gap, we offer a high-level overview of the emerging impact investment ecosystem.

Social impact investing ranges from angel investor networks like Investors Circle that typically invest in social impact start-ups (typically in the $50,000 to $3 million range) to social impact venture funds such as the Omidyar Network (launched by Pierre Omidyar, the founder of eBay) to impact focused private equity funds such as Bain Capital's $390 million Double Impact fund (led by former Massachusetts governor Devel Patrick). There are now hundreds of social impact investment funds, many tracked by the Global Impact Investing Network (GIIN) and conferences such as SoCap focused entirely on social impact investing.

An increasing number of leading philanthropies are also exploring social impact investing vehicles. This includes mission-related investing

(MRI), where foundations use their endowment (investment portfolio) to advance their philanthropic mission. Ford Foundation president, Darren Walker, recently announced that the foundation committed up to $1 billion for mission-related investments over the next ten years and many others are following. Program-related investing (PRIs) is a related investment approach that foundations pursue out of their grant-giving arm to diversify their impact portfolio, attract new social entrepreneurs, and stretch or recycle their capital. PRIs have a variety of structures, ranging from low-interest loans and loan guarantees to equity investments with impact-friendly terms.

Since many foundations and social impact investors have a low risk tolerance, especially when operating in new sectors, the fab ecosystem offers some lower risk investment opportunities. For example, fab-related services. One of the most important, and least sexy, parts of launching a fab lab involves the logistics of setting up a lab, procuring the necessary insurance and permits, maintaining and upgrading the equipment, ensuring safety, and all the other operational requirements of the day-to-day running of a lab. Addressing this challenge will also be a key part of cultivating an enabling fab ecosystem. For a long time, Sherry Lassiter and her tireless cohort of globe-trotting graduate students would take on this role, but they don't scale (indeed, the students graduate). This is why the Fab Foundation was established and it is well positioned to be a platform for these services. Others may see this as a low-risk investment opportunity since there is already a demand for these services and they can generate early revenue.

Academia

At the very core of an enabling fab ecosystem is the continued technological advancements ensuring Lass' Law maintains its exponential trajectory. The fundamental research that Neil and his colleagues are doing at the CBA, along with work by similar research organizations around the world, are essential to realizing this goal. The continuation of Lass' Law requires expanded support for basic digital fabrication research, as well as a strong iteration loop between the basic and applied research. Ensuring that the technology develops in a way that accelerates universal fab access and literacy requires integration of science and social science research.

The needed alignment across stakeholders in higher education is a considerable challenge under the current structures and cultures of universities, government labs, and private research labs. As we noted in

Chapter 4, there are incentives in these domains that drive conservative behaviors and undercut collective action. Neil had to establish CBA to do what he does—it didn't fit into any established academic categories. There is also a divide between research in the physical sciences and research in the social sciences. The opportunity is to build out a more robust ecosystem within and across the research communities, facilitating sharing of data, tools, and models, fostering inter-disciplinary collaboration, and pioneering new organizational and institutional models.

Russ Shilling, former executive director for STEM education at the US Department of Education and a former DARPA program officer, echoes this point: "Research that deals with inherently noisy data such as the behavioral sciences and education could benefit immensely from iterative research practices that erase the disconnects between basic and applied research." As Russ suggests, disconnects between basic and applied research are endemic to the social sciences. Bridging that gap—combining the work of path observers and path creators as well as scientists and social scientists will enable synergy across very diverse ecosystems that impact research priorities and how the research translates into early products and services.

Social Entrepreneurs

For fab social entrepreneurs who don't have the resources of governments, NGOs and leading philanthropies, there is a great opportunity to develop ideas that can propagate through a thoughtful mix of platforms, tools and practices. Currently there are large gaps in the fab ecosystem that ideally can be filled by impact-focused entrepreneurs that leverage a distributed network of empowered and empowering, deeply committed stakeholders as well as millions of engaged participations.

Many of the fab pioneers we have profiled in this book (any many more throughout the world) are working on projects that have the potential to emerge as globally transformative ecosystems. When these ecosystems, which share many of the same core objectives, emerge and intersect the seemingly insurmountable challenges, such as universal fab access, can become less overwhelming. For example, if Alycia Garmulewicz is able to develop an effective platform, with the associated tools and practices for sourcing and sharing local fab materials and Yogesh Kulkarni is able to do the same for distributed rural fab capability and Dina El-Zanfaly does the same for developing fab literacy—each of these ecosystems could significantly accelerate each other. These fab pioneers may choose to create new ecosystems, or they may chose to adapt and extend platforms developed by other fab pioneers.

There are a great many opportunities for social entrepreneurs to leverage new and emerging technologies that could be powerful elements of an enabling fab ecosystem. For example, consider augmented and virtual reality, as well as new location-aware technologies. Currently, much of the distributed fab mentoring is done through video conferencing, which has capacity issues. Augmented-reality solutions might serve as more scalable model over time. Similarly, there could be great value in creating simulated fab lab environments in virtual reality, where an individual with no fab access could get at least get a virtual experience. New location-aware technologies can also play a key role in a layered fab system. The multi-layered and distributed array of labs could be integrated into a location-aware mobile app that tracks fab capability and open access in any given location. If there is no capability locally, the app could also provide advice on how to launch a fab lab, a fab-enabled maker space, or a single-piece-of-equipment fab capability that could grow over time. This would be a valuable addition to an enabling ecosystem.

Two Personal Examples

Two specific aspects of contributing to the cultivation of an enabling fab ecosystem can be illustrated with respect to platforms for fab stakeholder alignment and fab literacy—domains that map onto the work that Joel and Alan do. These personal examples are just two among the many relevant examples of how people from very different disciplines can add value to the fab ecosystem.

Throughout this book we've highlighted numerous areas where independent but interdependent stakeholders need to align to accomplish goals around common interests, while navigating interests that may be in conflict. To help enable this alignment process, Joel has the goal of developing a platform for stakeholder alignment in complex ecosystems. The visualizations and analytics that he and his colleagues have developed are a step in that direction. Such a platform would support the specification of different types of stakeholders, at appropriate levels of granularity, and the identification of relevant interest (what is at stake). Then, such a platform would support the collection of data needed so that the architects of the platforms could see where there are points of alignment or misalignment. This would allow for pursuing low hanging fruit, avoiding dead ends, and planning for complex terrain that must, nevertheless, be navigated.

A good example of where such a platform could be a powerful rate accelerator would be around the aligning of stakeholders in support of common standards and protocols. This would, in turn, enable better collaboration across the current fragmented ecosystem of open source and proprietary hardware and software. Because the technology is constantly evolving, the use of the alignment platform would be an ongoing process. Another pivotal point of alignment would be among public and private investors in fab access. Not only might this avoid duplication and reveal synergies, but it would enable lateral alignment on metrics for demonstrated impact (a must-have for impact investors) and other enabling aspects of the investment ecosystem. In effect, a stakeholder alignment platform can be thought of as a platform multiplier—making the development of other needed platforms faster and more effective.

Another area where such a platform could be leveraged is to help align different levels of ambition and capability across diverse stakeholders. For example, some stakeholders are committed to being systems architects while others may or may not want to take on this role. They may, instead, prefer to be part of the empowered and empowering middle on a platform developed by someone else, which enables their aspirations. Ecosystems with platforms, tools, and process enable agency, but not everyone wants to have the same degree of agency with respect to a given ecosystem. Some will be architects, some the empowered and empowering middle, and some the distributed contributors. If there were a platform supporting stakeholder alignment, it would be, in effect, a meta platform, accelerating the cultivation of many other platforms in enabling ecosystems.

In advancing fab literacy, the Fab Academy is optimized for advanced students with access to full fab lab capabilities. There is a need for many types of formal and informal learning trajectories to cultivate fab literacy among people of various ages, skills and across different regions. These programs will need to be anchored in hands-on, project-based learning. In this area, there are insights that the emerging fab-based-learning community can benefit from those who have spent the last decade developing game-based-learning platforms, tools and practices. This is the area that Alan works in.

Both game-based and fab-based learning involve projects where people learn by doing. When effectively implemented both cultivate an environment for highly collaborative and passion-driven learning connected to real-world tools and communities of practice. Digital fabrication is hard, but difficult endeavors can still be engaging, empowering, and exhilarating—a challenge at the heart of good game design. Gamers are continually learning to master complex, technology-mediated experiences to accomplish goals they are personally invested in. Well designing games offer a delicate balance of challenge and rewards driving deep levels of engagement and enabling players to advance at their own pace, acquire critical knowledge just in time (versus just in case), and iterate according to copious feedback from the game, peers, mentors, and the community. This connection between game-based learning and fab literacy is at the core of the partnership across E-Line Media, the Fab Foundation, and CBA.

As more individuals and organizations pioneer new approaches to fab-based-learning there is a great opportunity to develop platforms, tools and practices for sharing everything from project ideas to teacher professional development to assessments. In this context, assessments are more about building a portfolio of individual and team-based projects, rather than about high-stakes testing often divorced from a real-world context. As Neil highlighted in Chapter 1, fab portfolios are already becoming an integral part of the Fab Academy accreditation process. This could be true for all levels of education.

Embedded Ecosystems

Like stakeholder alignment, cultivating ecosystems is not a one-time task, but a continuing accomplishment. Further, multiple embedded ecosystems are involved, each of which is in some ways independent and in some ways interdependent. Additionally, they operate as a layered system where the enabling platforms, tools, and practices function in similar, but distinct ways for individuals, communities, and global systems.

Although there is also the need for top-down catalyzing initiatives, we highlight cultivating ecosystems in the model since that is the key to enabling social systems to keep pace with accelerating technological change. These bottom-up and, crucially, middle-out change processes (employing platforms, tools, and processes) enables the propagation needed for a constructive co-evolution of society and technology. Further, it is possible also to then take into account other socio-technical ecosystems, such as those for artificial intelligence, robotics, biomedicine, and autonomous transportation that are relevant for the broader co-evolution of society and technology.

Indeed, it is even possible to envision the co-evolution with ecosystems in nature, as is illustrated by this observation by Beno Juarez, who grew up in the Amazonian jungles of Peru and is the architect of a floating fab lab on the Amazon that is oriented toward the biodiversity of the region. The lab is focusing on ways to print edible parts of a healthy diet with nutri-ents from the jungle and sensors to detect water contamination. As Beno notes, "the floating fab lab connects to a different life style in the jungle as compared to the city. Your scale changes in the jungle. You feel more con-nected to the planet." He adds that the Amazon floating fab lab is a "meet-ing of two rivers—technology and local culture—creating a new current."

Co-evolve: Technology and Society

The third digital revolution will not happen in isolation. The roadmap will unfold in a world of concurrent and complex economic, environmental, de-mographic, and sociopolitical change. It will influence and be influenced by broader economic trends such as technological unemployment and glo-balization, demographic trends such as the aging population and mass migration, environmental trends such as climate change and resource de-pletion, and political dynamics such as polarization and gridlock. Unlike the process of co-evolution in nature, the co-evolution of technology and society involves choices and actions. It is not just a matter of riding the wave, but involves influencing co-evolution's direction and impact—which is not easy to do. Co-evolution draws on all three parts of the model—observe, align, and cultivate—in a continuing cycle of change.

Among economic changes, the most pressing co-evolution dynamic lies at the intersection of everyone's ability to make (almost) anything and the growing job displacement associated with accelerating technol-ogies. In his 2015 book *Rise of the Robots*, Martin Ford argues that ad-vances in AI, automation, and robotics will increasingly replace both blue- and white-collar jobs and that these declines will, in turn, reduce tax revenues. This "soaring technological unemployment" and growing income and wealth inequality, when coupled with the accelerating costs of supporting an aging population and addressing the implications of climate change, could lead to a crippling downward spiral in society. In this scenario, the worst tendencies in human nature will be magnified by the growing desperation and anger of the displaced.

If this trajectory were to unfold, the third digital revolution could be a central part of the solution. Ford, like numerous other observers of ac-celerating job loss to technology, recommends some form of basic guar-anteed income, with carefully tuned incentives. Although this suggestion is emerging as a serious policy debate, one of the biggest issues with this solution is addressing the loss of dignity and purpose that comes with the

loss of a job. If, however, a basic guaranteed income also came with training in using digital fabrication for people to increasingly make what they
consume, there is opportunity for both economic and psychosocial benefit.
It is not a post-scarcity economy, but rather a post-salary economy.

In that case, it's a race. It will not be easy for digital fabrication to
match and exceed the pace of technological displacement. However,
if we make Lass' Law a reality and we see a billionfold improvement in
the capability and reach of digital fabrication, winning the race is possible. This will take aligned collective effort in a society that is right now
deeply divided. It will involve changes in the rules of the game—society's
institutions—a change that is both frightening and empowering. In the
eight aspirational scenarios, we saw a combination of urban and rural
self-sufficiency, community capability and learning, the use of local materials to address local solutions, cultural transformation, and entrepreneurial innovation. Together these scenarios serve as a counterpoint to
automation, job loss, and purchasing-power erosion.

Michael Piore and Chuck Sabel's 1984 analysis in *The Second Industrial Divide*, introduced in Chapter 4, suggests that in past divides,
the prior system doesn't disappear. With the rise of mass production, craft
production didn't disappear; it was needed to make the machines of mass
production, but it was no longer the dominant model. If distributed digital
fabrication does indeed grow at exponential rates, the role of multinational
corporations, service operations, financial markets, and entrepreneurial
initiatives will continue, but increasingly in the context of distributed digital fabrication.

In addition to the economic implications of the third digital revolution,
there will also be a needed co-evolution with the natural environment.
One of the biggest impacts of the third digital revolution will be on consumable materials and supply chains. If platforms emerge for the use of
local, environmentally friendly materials in fab labs, then there will be
major environmental gains. If not, then the exponential growth of fab labs
could be an environmental nightmare. Having billions of people tinkering
and iterating with environmentally unfriendly plastics (or other nonrenew-
able and scarce resources) is certainly not sustainable. For the health of
our planet, we need to work on creating an ecosystem that uses environmentally friendly consumables and reusable digital materials.

Another emerging implication of climate change (and political conflict)
is the growing crisis of displaced people and refugees. The UN High Commissioner for Refugees estimates that there are sixty-five million refugees,

asylum seekers, and internally displaced people around the world. Digital fabrication in refugee camps and temporary communities could offer multiple benefits—from modular, rapidly deployable shelter, furniture, and other basic needs to the building of technical skills and the fostering of collaborative cultures.

In the aspirational scenarios, the model that Blair Evans is pursuing for self-sufficiency in Detroit could easily transfer to refugee camps, where the ability to design and deploy low-cost solutions for hydroponics and housing could become a life-saving capability for refugees and displaced people. Moreover, individuals who build these capabilities in a refugee camp may be able to build portfolios for submission to a skills certification platform that eases their reentry back home or their emigration to a new community. Similarly, the displacement of some coastal communities because of climate change affords a unique opportunity for people looking to rebuild communities inland. We know that leaving homes and communities will be emotionally wrenching, but introducing the opportunity to start anew with enhanced capabilities may ease the transition. At a time when global displacement is growing, digital fabrication can be a new anchor for skills, capability, and sustainability.

With demographic change, aging is already part of the digital fabrication world in a few ways. The age profile of many fab labs is diverse—ranging from young children to retirees. Because many retirees are looking for enriching, social, and meaningful activities, fab labs could provide these older adults who have a lifetime of maker skills the opportunity to serve as mentors. (Conversely, the youth could develop important skills mentoring the retirees on the new digital technologies.) We already see numerous fab projects focused on mobility, physical assistance, and other projects useful for the baby boomer generation as it ages. Given that the number of retirees just in the United States is projected to increase from 43.1 million in 2012 to 83.7 million in 2050, there is urgency to work through the process of observation, alignment, and cultivation to get to co-evolution.

Consider what is possible in one sector of the economy and the associated communities hard hit during the last recession. Between 2007 and 2010, the US auto industry displaced more than two hundred thousand workers, many of whom took early retirements. Most of these individuals know their way around machine tools and CAD. There is a great opportunity to pair them with newly installed fab labs in their respective communities. We could then curate new operating methods (with enabling platforms) that would accomplish three things at once. First, for the retirees, we would be providing a meaningful, enriching postcareer experience—part of a rethinking of retirement that is happening more broadly in society. Second, these folks could apply the Mel King "learn to teach; teach to learn" process. King has high school students learn how to operate in a fab lab so that they can then teach junior high school students to do the same. With retirees, a similar approach would involve intergenerational co-creation and dialogue. Third, instead of being a burden on society, the declining economic sectors and retirees become central to community revitalization. The co-evolution is really a regeneration of society that spans generations.

The polarization and gridlock in politics and more broadly in society have been magnified by the first two digital revolutions. One digital cause of this polarization has been echo chambers, social media where people can listen and talk to people like themselves with fewer and fewer interactions with people who think differently. Digital media has also helped enable passionate networks of people to tear down governments, such as in what is referred to as the Arab Spring. But it is easier to tear down institutions than it is to build them back up, aligning the interests of diverse stakeholders. Into this mix come fab labs, which are typically open to anyone in the community. In contrast with virtual communities, it is common to see people of very different backgrounds and identities working side by side on projects in a fab lab. There is something about a tangible maker or fab project that fosters collaboration. In the Fiat Chrysler aspirational scenario, we introduce the social science concept of a boundary object—something that sits at the intersection of different communities and that enables the bridging across the boundaries. We said that fab labs are boundary objects, and it turns out that the projects taking place in the labs are themselves also boundary objects—small bridges within a larger bridge across communities.

By observing this bridging role of fab labs and fab projects, we see opportunities for alignment among people who are deeply troubled by the polarization in society. They can then cultivate the role of the fab lab by reaching out in communities with diversity in mind. The co-evolution will, however, be complex. A 2017 report titled "An Economy for the 99 Percent," by Oxfam, a British-based, international organization fighting poverty, observes that "from Nigeria to Bangladesh, from the U.K. to Brazil, people are fed up with feeling ignored by their political leaders, and millions are mobilizing to push for change." The report adds that "seven out of 10 people live in a country that has seen a rise in inequality in the last 30 years." In those settings, where income inequality is on the rise, fab labs could either provide alternative economic paths that temper the divisions or they could end up primarily serving those more fortunate and, as a result, deepen the divides.

In some countries, on the other hand, the middle class is growing and income inequality is shrinking. There, a new generation is—sometimes for the first time—experiencing the prospect of the next generation's doing much better. In these cases, digital fabrication is riding a different wave. The key is to foster co-evolution dynamics wherever collaboration-minded people enter society, from the fab lab crucible as an antidote to the tensions tearing apart some societies to a fab lab that reinforces class-mobility benefits in others.

In all these co-evolution scenarios, the process is the same. We begin with anticipation, alignment, and cultivation of enabling ecosystems to set the co-evolution dynamic in motion. The dynamic includes the narrow co-evolution between digital fabrication technology and the associated social systems, as well as the broader co-evolution between digital fabrication and broad societal challenges. The roadmap for digital fabrication will not be restricted to its own world of fab labs, but will influence and be influenced by broader economic environmental, demographic, and socio-political trends.

Organizations That Make Organizations and Institutions That Make Institutions

If machines can make machines, it is likely that organizations and institutions will need to co-evolve in analogous ways, which raises the interesting prospect of organizations making organizations and institutions making institutions. As shown in Chapter 3, the concepts of error correction, modularity, locality, and reversibility are what enable the machines to make machines (as well as the further exponential scaling to digital and programmable materials). Can we apply these concepts to social systems?

There is a mixed track record in applying analogies from science to social science, but here it is more than just parallel wording. When technology and markets shifted from craft to mass production, organizations and institutions changed as well. If technology and markets are oriented around digital fabrication, there will be a similar co-evolution. But this is not a deterministic argument around technology. We have choices to make about how this will happen, and there are constraints in the process. After all, the people in social systems have agency (and uniquely human biases and foibles). But this bug can also be a beneficial feature if the agency is understood and valued constructively.

One aspect of the science of digital—modular assembly and disassembly—is relatively familiar. Already, organizations born of the second digital revolution are organized around team-based work systems and project-based initiatives in which the needed talent assembles and disassembles with relative ease. We also see this modularity in the growing popularity of modular workspaces, where start-ups can easily rent space as they ebb and flow in size and hone their focus and strategy. This process of assembly and disassembly does take a specialized set of coordination skills, agile project management, and mitigation of harm during the change. Similarly, a wide variety of joint ventures and other forms of organizational collaboration allow for more macro forms of assembly and disassembly. At the institutional level in society, the growing use of consortia and other multi-stakeholder initiatives can serve as catalysts for change that would not have happened within existing institutional silos. These more flexible operating practices are not a perfect match for the way packets of information are assembled and disassembled, but they do have some similar properties. Moreover, the current experiences with assembly and disassembly in organizations and institutions are table stakes for the third digital revolution, which will require far greater assembly and disassembly as the pace of change accelerates.

The appeal of modularity for leaders is clear, though not always for individuals who may be working in the modules that have just been disassembled. David Weil, economist and former head of the US Department of Labor's Wage and Hour Division, writes of the "fissurization" of work— where careers have been decomposed into jobs that are now decomposed into projects, tasks, and contract relationships. He documents a dark side to work that becomes more modular: it is the workers more than the decision makers who are bearing the risk. At modularity's worst, organizations that make organizations, and institutions that make institutions, could achieve these objectives at even greater costs to individuals. At its best, such flexibility can accelerate innovation. For example, the Mondragon cooperative illustrates how the modular approach works well when there is a robust foundation for displaced workers to be supported as they are retrained and even financed if they have new business proposals.

The analogy to error correction may not be as intuitive as modularity. Yet, in many ways, error correction is the secret sauce of the first two digital revolutions—allowing for communication and computation at scale, without degradation of performance. Neil's key insight is that the same properties apply to digital fabrication. To some degree, there are some current analogues for organizations and institutions. For example, the principle of checks and balances in governance is a form of error correction at the institutional level. In industry, the reporting of near misses in manufacturing and airline safety is a form of error correction—identifying potential problems before they occur. The same principle applies in quality improvement. Such reporting is not easy to foster, given people's fear of blame or even simple embarrassment. When such reporting of near misses does occur and scales across entire enterprises or systems (such as air transportation), you begin to create a common culture of increased transparency, reduced blame, and shared responsibility for prevention. This is directly analogous to how the scaling of digital systems is enabled by error correction. The third digital revolution will require broader and more robust forms of error correction on the social side, enabling co-evolution with the accelerating technology.

ASSEMBLING ASSEMBLERS AND PANCAKE BREAKFASTS

The first two digital revolutions fundamentally changed how we live, learn, work, and play. Bits changed the destiny of atoms. The third digital revolution raises the stakes, because we will increasingly be able to manipulate and share bits *and* atoms. With accelerating advancements in the technology, we will be able to design reality, from food to furniture, tools to toys, crafts to computers, organizations to institutions.

Throughout this book, we have explored both the great benefits and the risks that come with this accelerating capability. But another question remains. Will these powerful new technologies make us happier?

The industrial revolution improved many people's lives, but it also created a mass-production culture that was soul-deadening. As Max Weber, one of the founders of modern sociology, wrote over a century ago, "it is horrible to think that the world could one day be filled with nothing but those little cogs, little men clinging to little jobs and striving towards bigger ones. . . . This passion for bureaucracy . . . is enough to drive one to despair."

The first two digital revolutions improved many lives, but also left many people feeling overwhelmed and longing for a simpler, less turbulent past. From digital-free holidays to digital detox programs to initiatives like the National Day of Unplugging, there has been a groundswell of people who feel as if their lives have become too saturated by digital technologies. We have even seen a resurgence of analog products and services like vinyl records, board games, and independent bookstores. Sherry Turkle eloquently described the phenomenon in her 2011 book *Alone Together*: "Technology offers us substitutes for connecting with each other face-to-face. We are offered robots and a whole world of machine-mediated relationships on networked devices. As we instant-message, email, text, and Twitter, technology redraws the boundaries between intimacy and solitude. . . . We recreate ourselves as online personae and give ourselves new bodies, homes, jobs, and romances. Yet, suddenly, in the half-light of virtual community, we may feel utterly alone."

As we hurl forward into a world with ever-more-compelling virtual and augmented-reality and other apps created by brilliant designers and behavior scientists to keep us staring at screens, finding a balance between time in the world of bits and time in the world of atoms could become increasingly difficult. The third digital revolution might just help realign this balance, because this digital revolution could help us leverage bits to spend more time in the world of atoms.

Making things is a deeply human behavior. Everyone needs to feel that their actions matter, that what they do makes a difference for themselves, their families, and the world. The more we make what we can consume, the greater our sense of self-worth. Community fab labs also tap into the deep human desires to make connections, to collaborate to tackle problems, and to accomplish goals together. As Dov Seidman says in Tom Friedman's *Thank You for Being Late*, "our ability to forge deep relationships—to love, to care, to hope, to trust, and to build voluntary communities based on shared values—is one of the most uniquely human capabilities we have. It is the single most important thing that differentiates us from nature and machines."

Fostering informal connections among people within and across fab labs is central to the fab culture. Haakon Karlsen tells the story of how, when he was co-designing the lab in the Lyngen Alps with Neil, He told him where the kitchen would go. Haakon says that Neil told him, "You can't have a kitchen in a fab lab." Years later, when Neil saw how the kitchen had become integral to the community lifeblood of the lab, Neil concluded, "You can't have a fab lab without a kitchen."

When we started writing this book, we began with the premise that digital technologies are exponentially accelerating but that individuals, organizations, and institutions develop linearly. Perhaps this is a feature, not a bug. Sure, we need to co-evolve to make sure that we shape the technologies so they don't shape in ways we will regret. But we must also retain the essence of what makes human—what gives life meaning.

Today, at age eighty-eight, Mel King, founder of the SETC fab lab in Boston, continues to champion how the ability to design and make (almost) anything is transformative. He concludes: "Fab labs are an imperative. It is a way of life. It is way of love. Making and creating . . . that is the bottom line."

Ultimately, the best advice we can give those who are thinking about cultivating the soul of the third digital revolution is to make sure that we don't lose the visceral exuberance and empowerment that is represented in these quotes from our survey of fab leaders summarizing in a word or phrase their view of fab labs and digital fabrication:

Dream, make and share
  —*Netherlands*
The path to inclusive industrialization
  —*Kenya*
If you can think it, do it
  —*Spain*
Materialize your ideas
  —*Colombia*
Misunderstood genius kid
  —*India*
Awesome chaos
  —*United Arab Emirates*
A space to revolutionize the power of the individual
  —*United States*
Fab labs should disappear and merge into society
  —*Chile*
Makes impossible things possible
  —*Panama*
A tool to change the world
  —*France*

These fab pioneers are, indeed, designing reality. Soon, everyone will be able to do so as well.

Conclusion

The year is 2017. Gas is $2.38 per gallon. Drake's *More Life* is topping the charts. Trump is in the Oval Office, Macron is in the Élysée, and Brexit has become official. *Star Wars: The Force Awakens* is leading the box office. The new ShopBot Desktop MAX debuts at $9,857. Epilog has introduced its new Fusion laser systems starting at $7,995, and the new Sindoh 3DWOX 3D printer comes out at $1,299.

You are sitting in a Starbucks enjoying a venti half-sweet, no-foam caramel macchiato. You happen to sit down next to Neil, Joel, and Alan Gershenfeld, who are having a lively debate about their new book, *Designing Reality*. At first the ideas they are debating seemed far-fetched. Billionfold improvements in digital fabrication performance? Anybody being able to make (almost) anything? Personal fabrication? You express skepticism. They tell you the signs are there if you know where to look. They then show you where to look.

The more you learn, the more you realize that this is indeed something you need to take seriously. The exact timing may be unclear, but you realize that you may have just been given an opportunity to see around the corner into a radically different future. Now you have to decide what to do with this knowledge. To help with that process, Neil, Alan, and Joel offer the following action steps.

GOALS AND RECOMMENDATIONS

The third digital revolution will not happen to us; we will make it happen. In this spirit, we offer a series of goals and recommendations all building on the roadmap and the model for predictive transformation. We start with suggestions for the sciences and social science, followed by recommendations for each stage of the projected roadmap.

Most of the world is not familiar with, or only vaguely aware of, digital fabrication. For many, digital fabrication is simply 3D printing. By recognizing that digital fabrication builds on and extends the underlying science that drove the first two digital revolutions, we will accelerate the necessary basic research during the third digital revolution: helping people understand both the full range of technologies for digital fabrication in the short term, as well as the emerging science of fabrication in the long term. Given this, our first overarching societal goal is the following:

Commit to advancing the emerging science of digital fabrication.

This goal might sound straightforward, but today, related work is fragmented across multiple institutional boundaries. Toward this goal, we have a number of supporting recommendations, including this one for scientists and researchers:

Integrate the development of digital communication and computation with that of digital fabrication.

The core thesis of this book is that technical and social systems need to proactively co-evolve, with the greatest influence occurring early on in the process. Thus, we have this recommendation for social scientists:

Recognize and support the capacity of the social sciences to proactively help shape the third digital revolution through advances in theory, tools, and methods.

Ultimately for effective co-evolution of the technical and social elements of the third digital revolution to emerge, each of the communities must embrace this process. For this to happen, we make this recommendation:

Establish formal and informal opportunities to bring together social scientists, technologists, storytellers, and other stakeholders to co-create integrated aspirations for the third digital revolution.

Given that Lass' Law is not a law of nature but will be driven by a combination of technological innovation and socially constructed forces, we also recommend this for all fab stakeholders:

Foster a shared sense of urgency, focus, and resources to drive continued exponential gains in digital fabrication performance and reach.

With these overarching goals and supporting recommendations in mind, we now turn to the four stages of the roadmap. Most of the goals and recommendations are targeted at the first stage—community fabrication—since this is foundational for all that follows. As the number of recommendations indicates, considerable foundational work is needed.

Community Fabrication

The technology for fab labs is here right now, and we have had ten years of developing the social systems to effectively run a fab lab. With this in mind, here is an overarching goal for the third digital revolution, starting with the community fab lab phase:

Make universal fab inclusion a global priority.

To help accomplish this goal, we offer the following recommendations for specific groups of stakeholders across the fab ecosystem.

Social-Impact Investors

Establish digital fabrication as a leading social-impact investment sector.

Engage leaders from the fab ecosystem to advise investors on opportunities for maximizing both social impact and financial returns.

Develop mechanisms for tracking the economic and social impacts of fab investments.

Government Leaders

Spend time in local fab labs to understand the impact on communities (and share what you learn).

Integrate digital fabrication technologies with existing digital divide initiatives.

*Step in where philanthropy and industry are not effective, ad-
dressing the most difficult rate limiters for creating public value
and mitigating harm.*

Philanthropists

*Focus grants, mission, and program-related investing on key fab
threshold issues, and support a fab enabling ecosystem.*

Seek synergies across public and private-sector fab funders.

*Embrace the opportunity to become the "Carnegie" of public fab
labs.*

Community Leaders

*Connect local fab labs in businesses, colleges and universities,
and community-based organizations.*

*Make fab labs in every town and school a basic assumption, like
public and school libraries.*

Stakeholders Across Sectors

*Track, document, and build on examples of fab labs and personal
fabrication enabling individuals and communities to become
more self-sufficient.*

*Integrate fab and maker pioneers, including those in education,
economic development, prosthetics, food, and security, into social-
impact organizations.*

For the current fab pioneers and entrepreneurs, we offer the following recommendations focused on common standards and shared principles:

*Fab labs, maker spaces, hacker spaces, and other similar facilities
should collaborate on common standards and shared principles,
while maintaining independent identities and approaches.*

Producers of digital fabrication equipment and software should also collaborate on common standards and shared principles, while maintaining independent identities and approaches.

The commons for digital fabrication should be supported, including the shared technological, organizational, and institutional infrastructure.

We also offer these recommendations focused on collaboration and knowledge sharing within and across communities:

Provide mechanisms to propagate effective practices within and across digital fabrication ecosystems, periodically mapping points of alignment and misalignment among stakeholders.

Continuously evolve the fab charter so that it can serve as a shared foundation for governance, operations, and growth— spanning stakeholders and ecosystems.

Build on small agreements around access, literacy, ecosystems, and risk toward societal grand bargains.

We make the following recommendations around creating an open and robust market ecosystem:

Cultivate marketplaces and distribution platforms that are effective, open, transparent, and aligned with the fab culture.

Continually scan for and support promising must-have products and services, drawing in capital to the sector.

Develop shared platforms for understanding, sourcing, and building with local, reusable, and environmentally friendly materials.

Design the work associated with digital fabrication so that it is characterized by fair treatment and constructive resolution of disputes.

Have a plan for when there is resistance to the democratization of digital fabrication by those whose interests are threatened.

People are, of course, at the heart of the fab ecosystem. Here are recommendations to increase the human capacity:

Identify, support, and expand the number of mentors and leaders within and across community fab labs and related facilities.

Recognize the need for individual and collective input in developing digital fabrication technologies.

And, with a nod to Haakon:

Every fab lab should have a kitchen.

Personal Fabrication

The second stage of the roadmap is comparable to the computer kits used by early adopters in the 1970s and 1980s. In personal fabrication, there is a concern that these millions of early personal fabrication adopters will be a select and fortunate group, with the vast majority of the planet left behind. With this in mind, we suggest this goal:

Foster the development of extensible, cost-effective, and intuitive tools for digital fabrication in the home and in small businesses.

To accomplish this goal, we need researchers and fab pioneers to develop examples of how individuals can have varying levels of fab capability in the home and small businesses, beyond a 3D printer. To help accomplish this goal we offer the following recommendations:

Partner with anthropologists and other social observers to follow early adopters in small businesses and the home, documenting emergent use patterns and identifying potential rate limiters and accelerators.

Develop protocols and standards so that industrially generated modular components that are hard to produce locally can be integrated into personal fabrication.

Track emerging economic models (barter, commercialization, open source, blockchain, etc.) to shape digital fabrication practices in ways that advance both sustainability and personal fulfillment.

Embrace the development of organizations that can make organizations, and institutions that can make institutions, to grow along with machines that can make machines.

Universal and Ubiquitous Fabrication

The third phase, universal fabrication, involves going from a million to the equivalent of a billion fab labs. The fourth phase, ubiquitous digital fabrication, leads to the equivalent of a trillion fab labs, when digital fabrication makes not just (almost) anything, but also (almost) everything. For all but the most technically oriented readers, these last two phases are very abstract. It is hard to visualize exactly how people will make things with digital materials. Even more challenging is to understand where people fit into the process of assemblers assembling assemblers. Ultimately, the underlying technologies, and the algorithms that drive these technologies, should be created with integrated humanistic values. Therefore, the broader population needs to advance goals such as the following:

Harness the power of popular media to create inviting, aspirational visions of preferred futures enabled by digital and programmable materials.

Build transparency into advanced digital fabrication technologies and algorithms.

Even with the best of intentions, however, there will be unintended consequences. Mechanisms for risk mitigation are essential. These are best addressed now, when there are fewer countervailing practices and entrenched interests.

*Bring risk-mitigation methods into the development of digital
and programmable materials now, when it is easiest to do so.*

Each of these summary recommendations builds on the others—the eco-systems needed for community fabrication continue to be important when it comes to personal fabrication, and the modular nature of personal fabrication lays the foundation for the universal and ubiquitous phases. All the phases depend on continued innovation of both technical and social systems. Many of these recommendations will not be easy. But we can collectively accomplish these goals if we are motivated to do so and have the enabling platforms and practices. We need to shape (and continually reshape) our organizations and institutions even as we live within them. To paraphrase Apollo 9 astronaut Rusty Schweickart, we are not passengers on the third digital revolution roadmap. We are the crew.

The authors. *Gladys and Walter Gershenfeld*

Epilogue

When we first proposed this book, the reaction from our agent (John Brockman) was, "Great idea! Lose the three authors." When it reached our editor-to-be (T. J. Kelleher), he said the same thing. We explained to both that the three-way collaboration wasn't incidental; it was essential to the story we wanted to tell.

The genesis of *Designing Reality* came in conversations we were having while we visited our mother, whose health was declining. She was no longer able to join in, but she enjoyed hearing our voices. As we talked, we realized that our careers, which had started out in very different directions, were colliding at the intersection of digital fabrication. The idea of turning that observation into a book literally began as a labor of love.

Writing the book proved to be more like a labor of labor, a microcosm of all the challenges and opportunities that we've outlined for the third digital revolution. We never realized just how dissimilar each of our thought processes were until we tried, and failed, to write a single narrative thread. Our second attempt, segregating each of our thoughts into a series of individual vignettes, succeeded in being so fragmented that T. J.'s charitable assessment was, "When you're driving down the wrong road, it's never too late to turn around and go back."

Only then did we appreciate the challenges to come in unifying our different approaches and perspectives. In fact, when we would tell people that we were writing a book together, the most common reaction was, "And you didn't kill each other?" Then we'd hear stories of painful and difficult sibling collaborations. Ultimately, though, our different perspectives and our ability to challenge one another enabled us to synthesize our observations from science, social science, and the humanities.

Neil's career began at Bell Labs, where he worked with lasers and particle accelerators to do experimental atomic and nuclear physics, basic research remote from any broader implications. But he ended up traveling to some of the most remote parts of the world to launch fab labs, and he helped create the Science and Entertainment Exchange, a National Academies office in Hollywood, to bring serious science into popular media.

Alan started out in Hollywood and in China working in the film industry. He was then recruited to help turn around video-game publisher Activision, where he ended up running its development studio. He went on to co-found E-Line Media to harness the power of games for learning and social impact, working with partners ranging from the Cook Inlet Tribal Council to the White House. From there, he has written articles for *Scientific American* and served as a principal investigator on National Science Foundation and DARPA grants, including teaming up with the CBA and Fab Foundation on games for fabrication.

Through Joel's career in facilitating labor-management cooperation and implementing high-performance work systems, he was appointed dean of the School of Labor and Employment Relations at the University of Illinois, president of the Labor and Employment Relations Association, and editor for the *Negotiation Journal* at the Program on Negotiation at the Harvard Law School. His work on stakeholder alignment for the National Science Foundation led him to study fab labs as part of the larger landscape of aligning institutions around grand societal challenges.

Even though fab labs are a point where we all meet, we think about them very differently. Neil practices ready, fire, aim—often trying things first in the spirit of rapid prototyping, without fully understanding their consequences and only later evaluating their implications. In writing here about Lass' Law, he had multiple formulations, seeing this flexible characterization as functional, something to test. Alan lives in the world of managing creative-impact projects that need to crystallize their vision and message; he insisted on consistency in describing Lass' Law for overall coherence. Joel, the social scientist, wanted to turn the whole Lass' Law thing into testable propositions, whereas Joel, the facilitator, just wanted everyone to be sufficiently well aligned to move the writing forward.

The seeds of negotiating our differences could be traced back to our family dinner table. We grew up in Philadelphia, but our parents took us on sabbaticals to live in Arizona, Jamaica, and England. Throughout our youth, our parents encouraged us to explore the world, pursue our passions, and make a positive difference in the lives of others. And to never be afraid of a bad pun (this book in three words: Lass is Moore).

The second question we would get when telling people about the book was, "What did your parents do?" Our parents, Walter and Gladys, were both labor arbitrators, professors of labor relations, and leaders of their national associations (Joel went into the family business). A regular topic of conversation at the dinner table was arbitration cases brought home by our parents. These cases required evaluating the evidence and arguing for interpretations. Getting a word in edgewise required skill, persistence, and a sense of humor. The moment of taking a breath by whoever was talking was an opening for someone else to take over. That's also what it was like writing this book.

We've come to appreciate that our difficulties and disagreements in working on the book reflect relative weaknesses in the worldviews of each of us—weaknesses that we've called each other on. We could easily write separately for followers of technology, popular media, and management; it has been much harder to merge our backgrounds to address the intersection of all three backgrounds at the same time. But that's where the third digital revolution will happen.

ACKNOWLEDGMENTS

As we started writing our acknowledgments, we realized that we were going to end up with roughly as many names of people who made essential contributions to this book as there are fab labs today (around a thousand). Rather than add that many names here, we offer our deepest gratitude to everyone mentioned throughout the book and the many others who have helped us develop and tell this story.

We've been fortunate to work with the publishing wizards at Basic Books and Brockman, Inc., the best in the business. And we're grateful for the thoughtful feedback from our many readers.

It's been said that *book* is a four-letter word for the way it takes over lives. Our collaboration in writing this has rested on the patience, support, curiosity, criticisms, and contributions of our families, for whom we're eternally grateful.

RESOURCES

We provide here background and references for the sources we've used in writing *Designing Reality*. They appear in the same order as the material appears in the chapters. The resource information is retrospective; we invite you to join us at the book's website, http://designingreality.org, for links to online content and updates to material in the book.

INTRODUCTION

Our face icons were drawn by Steve McCracken (who also designed the CBA and fab lab logos).

We've committed an editorial sin by writing *Lass' Law* rather than *Lass's Law*. We pronounce it "Lasses Law," but prefer the visual emphasis of *LL* rather than *sss*.

Gordon Moore's article, "Cramming More Components onto Integrated Circuits," *Electronics*, April 19, 1965, 8, features his original observations, the remarkable predictions he made, and his insights into how he connected the dots.

We quote President Obama on the Internet being a necessity, not a luxury. For coverage, see Krishnadev Calamur, "Broadband a 'Necessity,' Obama Says, As He Pushes FCC to Expand Access," *NPR*, January 14, 2015.

In discussing exponential change, we cite Ray Kurzweil, *The Singularity Is Near: When Humans Transcend Biology* (Penguin Books, 2005). Ray has been tracking exponential change in technologies with more depth and breadth than anyone we know.

Dale Dougherty is the creator of *Make* magazine and the Maker Faire. His most recent book provides a tour of the maker movement, *Free to Make: How the Maker Movement Is Changing Our Schools, Our Jobs, and Our Minds* (North Atlantic Books, 2016).

The work of Incite Focus on shelter, energy, and food is presented at the Incite Focus/ Center for Community Production/home page, www.incite -focus.org. The Fab City Pledge has a supporting website at http://fab.city. Both are referenced here and in a few other places in the book.

Neil Gershenfeld's *When Things Start to Think* (New York: Henry Holt and Company, 1999) presented what became known as the Internet of Things, and his *Fab: The Coming Revolution on Your Desktop: From Personal Computers to Personal Fabrication* (Basic Books, 2007) introduced the fab lab movement.

E-Line Media, founded by Alan Gershenfeld and Michael Angst, is a video-game publisher passionate about making great games that help players understand and shape the world. Learn more at the company's home page, http://elinemedia.com.

Jim Gee brought his expertise in literacy to digital fabrication in "Literacy: From Writing to Fabbing," November 11, 2014, www.jamespaulgee .com/archdisp.php?id=71&scateg=Linguistics

CHAPTER 1: HOW TO MAKE (ALMOST) ANYTHING

Neil developed the original CBA proposal (http://cba.mit.edu) with Joe Jacobson, Isaac Chuang, and Scott Manalis, advised by Marvin Minsky. The first How to Make (almost) Anything class (http://fab.cba.mit.edu/classes /MAS.863) was planned by them, along with Joe Paradiso and John DiFrancesco. The adventurous NSF program managers who agreed to the fab lab experiment were Kamal Abdali and Mita Desai. The first fab labs were developed and deployed by a group that included Sherry Lassiter, Bakhtiar Mikhak, Amon Millner, Caroline McEnnis, Amy Sun, and Manu Prakash.

Mel King provides a personal perspective on the Tent City story in *Chain of Change: Struggles for Black Community Development* (South End Press, 1981), which is the context for the first fab lab.

The origins of Metcalfe's law, which was apparently first stated by George Gilder in 1993, citing Robert Metcalfe, is summarized in Carl Shapiro and Hal R. Varian, *Information Rules* (Harvard Business Press, 1999).

The application examples are drawn from personal experience working in fab labs, the Fab Academy (http://fabacademy.org), and the annual gatherings of the fab lab network (http://fabevent.org).

Fab lab program information is available through the Fab Foundation (http://fabfoundation.org), Academany (http://academany.org), and community portal sites (https://www.fablabs.io).

Barcelona's original Fab City commitment (http://fab.city) was developed by Tomas Diez with Vicente Guallart (city architect), Xavier Trias (mayor), and Antoni Vives (deputy mayor).

US Rep. Bill Foster reports on the initial fab and maker legislation at "Foster Introduces Legislation to Support Next Generation of Makers, Innovators," press release, May 25, 2015, https://foster.house.gov /media-center/press-releases/foster-introduces-legislation-to-support -next-generation-of-makers.

The Global Humanitarian Lab (www.globalhumanitarianlab.org) was launched by David Ott (from the International Committee of the Red Cross) and Olivier Delarue (from the UN High Commissioner for Refugees).

CHAPTER 2: HOW TO (ALMOST) MAKE ANYTHING

Individuals quoted by name here, and in Chapters 4 and 6, were interviewed by Joel and Alan in January, February, March, and April 2017, as part of the research for the book.

The DARPA Social Science Initiative, "Accelerating Discovery with New Tools and Methods for Next Generation Social Science," was announced on March 4, 2016, www.darpa.mil/news-events/2016-03-04.

The graph on Internet use was constructed with data from Internet World Stats, International Data Corporation, and Nua Ltd.

Victoria Rideout and Vikki S. Katz's report focuses on digital access in low-income families: "Opportunity for All? Technology and Learning in Lower-Income Families," Joan Ganz Cooney Center, February 3, 2016, www.joanganzcooneycenter.org/publication/opportunity -for-all-technology-and-learning-in-lower-income-families.

The stakeholder alignment survey was distributed through the Fab Foundation email list to 11,355 recipients, of which 3,439 opened the document. The survey was approved for use by the Brandeis Institutional Review Board, which ensures proper treatment of the subjects of research (human and others). Complete responses were received from 178 individuals (resulting in a 5 percent response rate among those who opened the document). We had requests to translate the survey into multiple languages, which we were unable to do in the limited time we had available to us. This affects the response rate and the interpretation of the data. When asked to indicate their primary role, the participants gave the following mix of responses:

PRIMARY ROLE	PERCENTAGE
Local fab lab	48.0
Maker Space	9.1
Hacker Space	2.3
Educator promoting design and fabrication	18.9
Community organizers promoting design and fabrication	5.1
Public leaders supporting digital design and fabrication	5.1
Fab Academy	2.9
Fab Foundation	1.1
Producing and selling digital fabrication equipment	0.6
Digital materials manufacturing	1.1
Other	5.7

Most individuals had more than one primary role. When the respondents were asked to check all that apply, here was the distribution:

ALL ROLES (CHECK ALL THAT APPLY)	PERCENTAGE
Local fab lab	81.1
Maker Space	42.3
Maker Faire	31.4
Hacker Space	17.1
Educator promoting design and fabrication	66.3
Community organizers promoting design and fabrication	55.4
Public leaders supporting digital design and fabrication	37.1
Fab Academy	28.1
Fab Foundation	19.1
Producing and selling digital fabrication equipment	12.6
Digital materials manufacturing	21.7
Other	8.6

Note that this table includes Maker Faire, which was not a primary role in the above table. Most of these individuals have only been in their primary role for less than five years (54.7 percent), with most of the rest in their roles 5 to 10 years (32.6 percent) and the balance longer (12.8 percent).

Over three-quarters of the respondents were male (79.1 percent), with the balance female (20.3 percent) or intersex (0.6 percent). Slightly over half (50.3 percent) have a graduate degree (non-doctoral), and another 15.0 percent have a PhD. Some 18.5 percent have a bachelor's degree, and the balance (16.2 percent) have an associate's degree, some college, or a high school degree. The age distribution is remarkably even across age groupings, with 30.7 percent under the age of 35, 28.9 percent between 35 and 44, 24.9 percent between 45 and 54, and 15.6 percent over the age of 55.

The survey instrument was developed by the company Joel co-founded, WayMark Analytics, and the Michael Haberman generated the visualizations. These are termed *z-flowers* since what are called z-scores (a mathematical transformation to have a mean of zero and a standard deviation of 1) are sometimes used in the generation of the visualization. The print copy of the book includes z-flowers with shades of gray. The color versions (and additional z-flowers) can be viewed at http://waymark systems.org:8000/report/FAB2017#.

These visualizations make it possible to see every stakeholder's response, rather than just clusters of responses as might be represented in a histogram. Small numbers of respondents who are either very positive or very negative stand out in this format and are often essential to understand when seeking stakeholder alignment.

When we report that respondents found a given issue very important, we did so on the basis of responses that were a .7 or higher on an eleven-point numerical rating scale where 0 was "not important" and 1.0 was "very important." Similarly, when we reported that an issue was very difficult, it was ranked a .3 or lower on an eleven-point numerical rating scale, where 0 was "very difficult" and 1.0 was "very easy."

When viewing the visualizations, first look for the value on importance to assess the degree to which something is a must-have. Second, look for the degree of contrast between importance and ease, which indicates the gap to be addressed and the level of pain in the system (pain is high if something is very important, but hard to do). Third, look for the variation in the responses. If difficulty is bimodal, for example, it is likely that bringing up the bottom (those who find something difficult) will be more effective than enhancing things for people who do not find it difficult. There may also be bright spots or pain points highlighted in the visualization that need further investigation. Altogether, these visualizations are designed to provide increased situational awareness and prompts for further action.

Kate Pickett and Richard Wilkinson document the benefits of equality in *The Spirit Level: Why More Equal Societies Almost Always Do Better* (Bloomsbury Press, 2009),

Gui Cavalcanti attempts to clarify terminology and culture in "Is It a Hackerspace, Makerspace, TechShop, or FabLab?," *Make* magazine, May 22, 2013. This topic would benefit from further dialogue.

The Freestyle Chess story was taken from Andrew McAfee and Erik Brynjolfsson's *The Second Machine Age: Work, Progress, and Prosperity in a Time of Brilliant Technologies* (W. W. Norton & Company, 2016).

Mariana Mazzucato's *The Entrepreneurial State: Debunking Public vs. Private Sector Myths* (Anthem Books, 2015) identifies the key role of public-funded research in driving private-sector innovation and growth.

The sample fab lab budget was provided by the Fab Foundation.

The Fab City white paper "Locally Productive, Globally Connected Self-Sufficient Cities" is available at http://fab.city/whitepaper.pdf.

For more information on alternative approaches to risk assessment, see Nancy Leveson et al., "Systems Approaches to Safety: NASA and the Space Shuttle Disasters," in *Organization at the Limit*, ed. William Starbuck and Moshe Farjoun (Wiley-Blackwell, 2005).

"The Science Behind Shaping Player Behavior in Online Games," Jeffrey Lin's talk at Game Developers Conference, 2013, outlines Riot Games' efforts to reduce toxic behavior in the *League of Legends* game (http://gdcvault.com/play/1017940/The-Science-Behind-Shaping-Player). Brendan Maher's "Can a Video Game Company Tame Toxic Behaviour?," *Nature*, March 30, 2016, also covers Riot Games' efforts to reduce toxic behavior.

US Rep. Bill Foster's proposed National Fab Lab Network Act of 2015 (H.R. 1622) continues in discussion in the US Congress.

More information on the UN Sustainable Development Goals is available at www.un.org/sustainabledevelopment/sustainable-development-goals.

CHAPTER 3: THE SCIENCE

The Moore's Law data was taken from Gordon Moore's original paper "Cramming More Components onto Integrated Circuits," *Electronics* 38, no. 8 (1965), and then from Intel's mainline chip specifications (http://ark.intel.com).

The Lass' Law data came from CBA's internal records, the wiki started by Frosti Gíslason (http://wiki.fablab.is), and the portal started by Tomas Diez (https://www.fablabs.io).

Shannon's original reference for reliable communication was C. E. Shannon, "A Mathematical Theory of Communication," *Bell System Technical Journal* 27 (1948): 379–423, and von Neumann's original reference for reliable computation was John von Neumann, "Probabilistic Logics and the Synthesis of Reliable Organisms from Unreliable Components," *Automata Studies* 34 (1956): 43–98.

The discussion of digitization in biology drew on the work of, and conversations with, George Church, Joe Jacobson, Pete Carr, Tom Knight, Erik Winfree, Paul Rothemund, Erez Lieberman Aiden, Alicia Jackson, John Glass, and Elizabeth Strychalski.

The discussion of digitization in physics drew on the work of, and discussions with, Charles Bennett, Rolf Landauer, David DiVincenzo, Nabil Amer, Scott Kirkpatrick, Geoff Grinstein, Ed Fredkin, Tommaso Toffoli, Norman Margolus, Seth Lloyd, Isaac Chuang, Jean-Jacques Quisquater, and Wojciech Zurek.

The discussion of digitization in intelligence drew on the work of, and discussions with, John Doyle, Raff D'Andrea, Pablo Parrilo, Ben Recht, Andreas Weigend, Terry Sejnowski, John Hopfield, and Sebastian Seung.

CHAPTER 4: THE SOCIAL SCIENCE

The presentation of the reactive roots of the social sciences draws on discussions with Scott Cooper, Dan Cornfield, Ton Kochan, Bob McKersie, and members of the Stakeholder Alignment Collaborative, including Karen Baker, Nick Berente, Pat Canavan, Gabe Gershenfeld, Brandon Grant, John Leslie King, Christine Kirkpatrick, Barbara Lawrence, Chris Lenhardt, Peter Levin, Spenser Lewis, Matt Mayernik, Charlie Mcelroy, Barbara Mittleman, Namchul Shin, Shelly Stall, and Susan Winter.

In *The Protestant Ethic and the Spirit of Capitalism* (George Allen & Unwin Ltd., 1905; trans. by Talcott Parsons, 1930), Max Weber demonstrated how religious conceptions of work, particularly Calvinist concepts, conferred legitimacy on a capitalist system.

Émile Durkheim's *The Division of Labor in Society* (1893) was based on his doctoral dissertation and traced the impact of the industrial revolution on class structure in society.

Octavia Hill, in *Our Common Land (and other short essays)* (Macmillan, 1877), focused on visiting the poor, the importance of charity, the value of open spaces, and related topics.

Jane Addams's 1892 essay "The Subjective Value of a Social Settlement" focused on Hull House in Chicago and was one of many essays she wrote on homes for the poor.

The International Union of Psychological Science traces its roots to the first international congress in 1885 in www.iupsys.net/about/history/index.html. The founding of the American Psychological Association is documented at www.apa.org/about/apa/archives/apa-history.aspx

Sidney Webb and Beatrice Webb, *Industrial Democracy* (Longmans, Green & Co., 1897), focused on collective bargaining as the foundation for "industrial democracy," a new term introduced into the social sciences by the Webbs.

John R. Commons coauthored the four-volume *History of Labor in the United States* (Macmillan, 1918–1935). It is part of a body of work connecting changes in institutional arrangements, such as the structure of unions, with changes in the structure of markets, such as happened with the impact of the Erie Canal on commerce.

Mary Wollstonecraft Shelley's *Frankenstein: or, The Modern Prometheus* (Lackington, Hughes, Harding, Mavor, & Jones, 1818) is one of the first modern examples of science fiction.

Charles Dickens's *Hard Times* (Bradbury and Evans, 1854) is a critique of the utilitarian philosophy that accepted the social order associated with the industrial revolution.

Upton Sinclair's *The Jungle* (Doubleday, Jabber & Company, 1906) illustrates the power of fiction to influence policy, such as what happened with legislation on the meatpacking industry. In later years, John Steinbeck's *The Grapes of Wrath* (Viking Press, 1939) and other writers continued the tradition of social commentary through fiction.

Path dependency as a concept emerged in economics with efforts to account for new technologies. See, for example, Paul A. David, "Path Dependence: Putting the Past into the Future of Economics," working paper, Institute for Mathematical Studies in the Social Sciences Technical Report 533, Stanford University, November 1988; and W. Brian Arthur, "Self-Reinforcing Mechanisms in Economics," in *The Economy as an Evolving Complex System: The Proceedings of the Global Economy Workshop, Held September, 1987 in Santa Fe, New Mexico*, ed. Philip W. Anderson, Kenneth J. Arrow, David Pines (Addison-Wesley, 1988).

Robert Michels's *Political Parties: A Sociological Study of the Oligarchical Tendencies of Modern Democracy* (Hearst, 1911) focuses on trade unions and political parties, both mission-driven institutional arrangements that focused first on continued existence and only after that on mission.

Thomas Kuhn's *The Structure of Scientific Revolutions* (University of Chicago Press, 1962) documents the conservative forces in institutions that constrain change, even in the evidence-based world of science.

Adam Smith's *An Inquiry into the Nature and Causes of the Wealth of Nations* (London: W. Strahan, 1776) has shaped thinking about markets that were just beginning to emerge in ways that contrasted with feudalism. Note that the original handwritten manuscript (with cross-outs and notes) is in the rare-books collection of the University of Illinois at Urbana-Champaign.

Karl Marx's *The Communist Manifesto* (1848) calls for the class struggle against the owners of the means of production. In this work and in jointly authored works with Friedrich Engels, there is surprisingly little on what it would mean for workers to own the means of production, in contrast with the ownership class. Increased individual and community ownership of the means of production is, of course, a prospect that is central to this book.

Frederick Taylor's *The Principles of Scientific Management* (New York: Harper and Brothers, 1911) was reprinted many times as a central reference for Taylor Society chapters around the country and is foundational for the field of industrial engineering.

Fred Emery and Merrelyn Emery wrote *A Choice of Futures* (Martinus Nijhoff Social Sciences Division, 1976) as part of a series of monographs on the quality of working life.

Stewart Brand's 1968 launch of the *Whole Earth Catalog* led to many editions, a *Whole Earth Epilogue*, and, beginning in 1974 (the year Joel graduated from high school), the launch of the *CoEvolution Quarterly*.

John Markoff's *What the Dormouse Said: How the Sixties Counterculture Shaped the Personal Computer Industry* (Viking Penguin, 2005) connects the rise of digital computation with countercultural forces in society.

Stewart Brand and J. Baldwin edited *Soft-Tech: A Co-Evolution Book* (Penguin, 1978), which is designed with the same look and feel as the *Whole Earth Catalog*, but with a focus on technology, broadly defined.

Alvin Toffler's *Future Shock* (Random House, 1970) is based on an article he wrote: "The Future As a Way of Life," in *Horizon* magazine.

Michael Piore and Charles Sabel's *The Second Industrial Divide: Possibilities for Prosperity* (Basic Books, 1984) connects the dots between changing computer technologies used in manufacturing and changing markets.

Ray Kurzweil's *The Age of Intelligent Machines* (MIT Press, 1992) is a precursor to his *The Age of Spiritual Machines: When Computers Exceed Human Intelligence* (Viking Press, 1999) and his *The Singularity Is Near: When Humans Transcend Biology* (Viking Press, 2005), all of which trace exponential rates of change in communication and computation technologies and the implications for core assumptions about human existence.

Arthur C. Clark's *2001: A Space Odyssey* (New American Library, 1968) was released concurrently with the movie, which he developed jointly with Stanley Kubrick.

When William Shatner visited Neil's lab to discuss the science of Star Trek while he was writing *I'm Working on That: A Trek from Science Fiction to Science Fact*, he and Neil agreed that they couldn't tell who was imitating whom.

William Gibson's "Burning Chrome," in *Omni* (July 1982), introduced the term "cyberspace," which was further elaborated in his *Neuromancer* (Ace, 1984).

Note that Joel finds inspiration in Robert Heinlein's concept of a fair witness in Heinlein's novel *A Stranger in a Strange Land* (Putnam, 1961)—a science-fiction idea that is desperately needed in society.

The World Bank's "2016 World Development Report on Digital Dividends" is available at www.worldbank.org/en/publication/wdr2016.

Gamestar Mechanic is a game platform and community that enables youth to design their own original video games. *Gamestar* is published by E-Line Media and was developed in partnership with the Institute of Play and the MacArthur Foundation. Since its launch, over one million original games have been published by youth player/creators. Learn more at https://gamestarmechanic.com.

Douglas McGregor's *The Human Side of Enterprise* (McGraw Hill, 1960) was reproduced in an annotated edition in 2006 with Joel as the editor. It is among the most influential management books ever published.

Joel Cutcher-Gershenfeld, Dan Brooks, and Martin Mulloy document fifty-six pivotal events in *Inside the Ford-UAW Transformation: Pivotal Events in Valuing Work and Delivering Results* (MIT Press, 2014), which identified the key role of changing deeply embedded operating assumptions to enable culture change. This, in turn, builds on Ed Schein's *Organizational Culture and Leadership* (Jossey-Bass, 1985).

A window into the culture of the Digital Equipment Corporation is provided by Guidon Kunda in *Engineering Culture: Control and Commitment in a High-Tech Corporation* (Temple University Press, 2000).

William Bridges's *Managing Transitions: Making the Most of Change* (Da Capo Press and Perseus Books, 1991) has recently been released in a special twenty-fifth anniversary edition, indicating its long-standing value.

MIT chancellor Cynthia Barnhart and medical director William Kettyle announced the MindHandHeart Initiative in Kath Xu, "Mind-HandHeart Expected to Improve Access to Resources," *(MIT) Tech*, September 3, 2015, http://tech.mit.edu/V135/N20/mindhandheart.html. For *Car Talk* host Ray Magliozzi's comments on the MIT seal, see Peter Dizikes, "Moving Spectacle: MIT Marks 100 Years in Cambridge with 'Crossing the Charles' Parade and Evening Celebration," *MIT News*, May 9, 2016, at http://news.mit.edu/2016/crossing-charles-moving-day-parade -competition-0509. The history of the MIT Seal is at http://libraries.mit .edu/mithistory/institute/seal-of-the-massachusetts-institute-of-technology. For recent commentary on head, heart, and hands at MIT, see coverage of Megan Smith's commencement address in 2015, http://news.mit.edu/2015 /commencement-day-0605.

Robert Owen's attempt to create a new society was not sustainable, partly because of what Owen's oldest son, Robert Dale Owen, described as "a heterogeneous collection of radicals, enthusiastic devotees to principle, honest latitudinarians, and lazy theorists, with a sprinkling of unprincipled sharpers thrown in," who were drawn to the community. An interesting legacy is that Owen's oldest son entered Indiana politics, where he championed legislation giving widows and married women control over property and promoted free schools in the state. He then represented Indiana in the US Congress, where he wrote the legislation establishing the Smithsonian Institution.

Mondragon is the subject of hundreds of books and scholarly articles. More information on the cooperative is at www.mondragon-corporation .com/en.

Remake Learning has become a benchmark for regional learning ecosystem transformation. More information on Remake Learning is at http://remakelearning.org. The Remake Learning Playbook, a field guide of ideas and resources for building innovation networks for teaching and learning, was created by the Sprout Fund; see http://remakelearning.org /playbook.

CHAPTER 5: THE ROADMAP

The roadmap presented in this chapter grew out of an event that CBA and the White House Office of Science and Technology Policy cohosted on the science of digital fabrication on March 7, 2013 (http://cba.mit.edu /events/13.03.scifab).

The fab lab inventory (for the community fabrication stage) is available at http://fab.cba.mit.edu/about/fab/inv.html.

William Gibson made his statement on the future being here now in a radio interview in "The Science in Science Fiction," Talk of the Nation, NPR, November 30, 1999.

The transition from rapid prototyping to the rapid prototyping of rapid prototyping (for the personal fabrication stage) is presented in Nadya Peek's "Making Machines That Make: Object-Oriented Hardware Meets Object-Oriented Software," PhD dissertation, Massachusetts Institute of Technology, Cambridge, MA, 2016.

The source for the robotic builder is Benjamin Jenett and Kenneth Cheung, "BILL-E: Robotic Platform for Locomotion and Manipulation of Lightweight Space Structures," 25th AIAA/AHS Adaptive Structures Conference, AIAA SciTech Forum, Grapevine, Texas, January 13, 2017 (AIAA 2017–1876).

The transition from analog to digital materials (for the universal fabrication stage) is presented in Kenneth Cheung's "Digital Cellular Solids: Reconfigurable Composite Materials," PhD dissertation, Massachusetts Institute of Technology, Cambridge, MA, 2012, and William Langford's "Electronic Digital Materials," master's thesis, Massachusetts Institute of Technology, Cambridge, MA, 2014.

The transition from assembly to self-assembly (for the ubiquitous fabrication stage) was introduced by John von Neumann and A. Burks, *Theory of Self-Replicating Automata* (University of Illinois Press, 1966), and the design tools are discussed in Amanda Ghassaei's "Rapid Design and Simulation of Functional Digital Materials," master's thesis, Massachusetts Institute of Technology, Cambridge, MA, 2016.

CHAPTER 6: THE OPPORTUNITY

Alvin Toffler's *Future Shock* (Random House, 1970), which we have noted in various places, is prescient in its appreciation for exponential rates of change.

Carl Sagan's quote is from his "Why We Need to Understand Science," *Skeptical Inquirer* 14 (spring 1990): page 3.

More information on Alex McDowell's World Building Institute at the University of Southern California is at http://worldbuilding.institute.

McDowell's quote is in Evan Atherton and Maurice Conti, eds., *Four: A Collection of Short Stories Exploring the Future of Design, Technology, and Us* (Autodesk Applied Research Lab, 2016).

The source for the bridge image is Benjamin Jenett, Daniel Cellucci, Christine Gregg, Kenneth Cheung, "Meso-Scale Digital Materials: Modular, Reconfigurable, Lattice-Based Structures," ASME International Manufacturing Science and Engineering Conference, 2016; and the source for the wing image is Benjamin Jenett, Sam Calisch, Daniel Cellucci, Nick Cramer, Neil Gershenfeld, Sean Swei, and Kenneth Cheung, "Digital Morphing Wing: Active Wing Shaping Concept Using Composite Lattice-Based Cellular Structures," *Soft Robotics*, March 2017, 4 (1).

Kurt Lewin's change model is in his *Field Theory in Social Science: Selected Theoretical Papers* (Harper and Brothers, 1951).

John Paul Kotter's model for leading change was first published in his "Leading Change: Why Transformation Efforts Fail," *Harvard Business Review*, March–April 1995, 59–67.

William Bridges's model for managing transitions is in his *Managing Transitions: Making the Most of Change* (Da Capo Press and Perseus Books, 1991), as noted earlier.

W. Edwards Deming's plan, do, check, and adjust model for continuous change was first published in his *Out of the Crisis* (MIT Press, 1986), page 88, as plan, do, study, act.

Everett Rogers's *The Diffusion of Innovation* (Free Press, 1962) located technology in the context of communication systems.

Geoffrey Moore's *Crossing the Chasm: Marketing and Selling High-Tech Goods to Mainstream Customers* (Harper Collins, 1991) is a touchstone for technology entrepreneurs making the transition from start-up to established enterprise.

Clayton Christensen's *The Innovator's Dilemma* (Harvard Business School, 1997) has spawned a small army of scholars challenging his initial estimates on disruption in various industries while celebrating his core insights.

NASA's "technology readiness levels" are summarized at Thuy Mai, ed., "Technology Readiness Level," NASA, last updated July 31, 2015, www .nasa.gov/directorates/heo/scan/engineering/technology/txt_accordion 1.html.

US Department of Defense's "technology readiness assessment" is summarized at Assistant Secretary of Defense for Research and Engineering, "Technology Readiness Assessment (TRA) Guidance," US Department of Defense, April 2011, www.acq.osd.mil/chieftechnologist /publications/docs/TRA2011.pdf.

Stewart Brand's "Self-Driving Genes Are Coming," his answer to a survey asking, "2016: What Do You Consider the Most Interesting Recent [Scientific] News? What Makes It Important?," appears in *Edge* (blog), edited by John Brockman (www.edge.org/contributors/what-do-you-consider -the-most-interesting-recent-scientific-news-what-makes-it).

Joi Ito and Jeff Howe's *Whiplash: How to Survive Our Faster Future* (Grand Central Publishing, 2016) offers nine principles for navigating a world of accelerating change.

The visualization of the four stages was constructed by beginning with the current rate of change with new fab labs, using Neil's data from Chapter 2. For the curve with the rate of change for community fabrication, an annual rate of change of 1.5 was used (doubling every year and a half), which was decelerated to 1.4 in 2020 and then progressively slower to level out around 2030. The same curve was then duplicated, beginning with a five-year lag for personal fabrication and a leveling off that begins around 2035. For the final two curves, lags of an additional five years were used for each, and the leveling off was projected much further out—at 2045 for digital materials and off the chart for programmable materials. Because this is an illustration rather than a charting of actual data, the format is different from the data charts used in earlier chapters. The full set of data is available on request from the authors.

Alvin Toffler's *Future Shock* (Random House, 1970) is also cited earlier.

Mihaly Csikszentmihalyi's *Good Business: Leadership, Flow, and the Making of Meaning* (Penguin Books, 2003) explores how the concept of flow applies to business and society.

Never Alone (Kisima Innitchuna) is a commercial video game developed through an inclusive development process by E-Line Media and pioneering Alaska Native elders, writers, and storytellers. Released in November 2014, it has been selected for over seventy-five best-games lists in

2014 and has won multiple awards, including a BAFTA (British Academy of Film and Television Arts) award and Game of the Year at Games for Change. Learn more at http://neveralonegame.com.

Coverage of the Ford Foundation's commitment of a billion dollars from endowment to mission-related investments is at Ford Foundation, "Ford Foundation Commits $1 Billion from Endowment to Mission-Related Investments," April 5, 2017, www.fordfoundation.org/the-latest/news /ford-foundation-commits-1-billion-from-endowment-to-mission-related -investments.

The Heron Foundation's announcement that all $273 million of its endowment was going to mission-related investing is Heron, "Heron's Journey," no date, at www.heron.org/enterprise#2011.

More information on the Investors' Circle social impact investing network is at www.investorscircle.net.

Andrew Carnegie's *Autobiography of Andrew Carnegie* (Houghton Mifflin, 1920), was published one year after his death.

The Gates-Buffet Giving Pledge is summarized at https://giving pledge.org.

See the World Bank, "Global Connect Initiative," https://share.america .gov/globalconnect, for a summary.

The ConnectED, FCC E-Rate Program is summarized in Federal Communications Commission, "E-Rate: Schools & Libraries USF Program," last updated April 28, 2017, www.fcc.gov/general/e-rate-schools-libraries -usf-program.

The concept of a layered system can be further understood with an example from the doctoral seminar in engineering systems that Joel cofounded at MIT in 2002. In comparing the U.S. and British health care systems, the U.S. system has been described as a two-layered system of doctor's offices and hospitals. By contrast, the British system has been described as a three-layered system of nurse-led clinics, doctor's offices, and hospitals. The U.S. system does have some nurse-led clinics, but they are not integral to the design. Layers are interconnected, but they also function independently, which adds robustness (and complexity) to the system.

The American Library Association defines digital literacy in Marijke Visser, "Digital Literacy Definition," *ALA Connect*, September 14, 2012, http://connect.ala.org/node/181197.

Martin Ford's *Rise of the Robots: Technology and the Threat of a Jobless Future* (Basic Books, 2015) explores the implications of accelerating technologies such as AI and robotics on the future of work and society.

Michael Piore and Charles Sabel's *The Second Industrial Divide: Possibilities for Prosperity* (Basic Books, 1984) was also cited earlier.

The UN High Commissioner for Refugees estimates that there are sixty-five million refugees at Adrian Edwards, "Global Forced Displacement Hits Record High," UNHCR, June 20, 2016, www.unhcr.org/en-us/news/latest/2016/6/5763b65a4/global-forced-displacement-hits-record-high.html.

Retirees projections are in *Current Population Reports* by Jennifer M. Ortman, Victoria A. Velkoff, and Howard Hogan, May 2014), P25-1140.

For background on the displacement of auto workers, see Joel Cutcher-Gershenfeld, Dan Brooks, and Martin Mulloy, *Inside the Ford-UAW Transformation: Pivotal Events in Valuing Work and Delivering Results* (MIT Press, 2015).

The Oxfam report on income and wealth inequality, "An Economy for the 99 Percent," January 2017, www.oxfam.org/en/research/economy-99, is one of a growing number of studies connecting technology, economics, and inequality.

The Max Weber quote is from 1909 and is cited by Reinhard Bendix, *Max Weber: An Intellectual Portrait* (Doubleday and Company, 1960), page 464. It reflects the visceral quality underlying his observations on bureaucracy.

David Weil's *Fissurization: Why Work Became So Bad for So Many and What Can Be Done to Improve It* (Harvard University Press, 2014) is at the center of current policy debates and is based on his leadership at the US Department of Labor around overtime rules and other policies focused on fair pay.

In *Alone Together: Why We Expect More from Technology and Less from Each Other* (Basic Books, 2011), Sherry Turkle explores the deep human connections and disconnects associated with digital technology.

The quote in the text is from Tom Friedman's *Thank You for Being Late: An Optimist's Guide to Thriving in the Age of Accelerations* (Farrar, Straus and Giroux, 2016).

The concluding quotes are from our survey of fab leaders who were responding to our request for "a phrase or metaphor to summarize your current view of fab labs and other aspects of digital fabrication." We selected what we thought were the most evocative quotes and only then discovered that they were from a wonderfully diverse mix of countries.

INDEX

ABOUT THE AUTHORS

© John Werner

NEIL GERSHENFELD has been called the intellectual father of the maker movement. He leads MIT's pioneering Center for Bits and Atoms and is the founder of the global network of community fab labs that has grown to more than a thousand sites. His earlier books *When Things Start to Think* and *Fab* presented what became known as the Internet of Things and the maker movement long before those terms became everyday expressions. He lives in Cambridge, Massachusetts.

© Ethan Gershenfeld

ALAN GERSHENFELD is a pioneer in harnessing the power of digital media for learning and social impact. As a former studio head at Activision, former chairman of Games for Change, and co-founder/president of E-Line Media, he has helped bring the power of games and digital media to engage and empower millions of youth and young adults. E-Line is currently working with the Center for Bits and Atoms and the Fab Foundation on a game to fire the imagination of a generation around the future of digital fabrication, with research funding from DARPA. Alan lives in Scottsdale, Arizona.

© Bethany Romano,

JOEL CUTCHER-GERSHENFELD is a world leader in workplace transformation and institutional change, with a client list ranging from Ford and the United Auto Workers to Australia's Fair Work Commision. He is professor in the Heller School for Social Policy and Management at Brandeis University and serves as editor of the *Negotiation Journal*, published by the Program on Negotiation at Harvard Law School. Joel is past president of the Labor and Employment Relations Association. He led the first stakeholder alignment map across the US fab lab network and co-founded the Champaign-Urbana Community Fab Lab. He lives in Waltham, Massachusetts.